STUDENT ATLAS of ENVIRONMENTAL ISSUES

D0166410

STUDENT ATLAS of ENVIRONMENTAL ISSUES

John L. Allen
University of Connecticut

Dushkin/McGraw·Hill

A Division of The McGraw·Hill Companies

Book Team

Vice President & Publisher   *Jeffrey L. Hahn*
Managing Editor   *John S. L. Holland*
Production Manager   *Brenda S. Filley*
Editor   *Pamela Carley*
Copy Editor   *Joshua Safran*
Designer   *Charles Vitelli*
Typesetting Supervisor   *Juliana Arbo*
Proofreader   *Diane Barker*
Graphics   *Shawn Callahan*

Dushkin/McGraw·Hill

A Division of The McGraw·Hill Companies

The credit section for this book begins on page 119 and is considered an extension of the copyright page.

Cover design   *Shawn Callahan*
Cover photo courtesy of NASA

Copyright © 1997 The McGraw-Hill Companies, Inc.
All rights reserved

Library of Congress Catalog Card Number: 97-65767

ISBN 0-697-35620-4

No part of this publication may be reproduced, stored in a retrieval system, or transmitted, in any form or by any means, electronic, mechanical, photocopying, recording, or otherwise, without the prior written permission of the publisher.

Printed in the United States of America

10   9   8   7   6   5   4   3   2   1

A Note to the Student

The quality of the global environment and the impact made by human activities upon that environment are among the greatest challenges of our time. In spite of extraordinary progress during the last 20 years in a number of environmental areas—particularly within the more developed world—vast areas of the world are subject to enormous stresses brought on by huge and growing human populations, by runaway land surface modification (such as deforestation), and by high rates of industrialization. In many areas of the developing world, pollution of soil, water, and air is proceeding at unparalleled rates. In many of those same areas, the use of increasingly marginal fragile lands—too dry, wet, or steep for stable and sustained land use—produces impacts upon the biosphere that are approaching the irreversible.

As the industrialized countries of the former Soviet bloc have become more open to Western scientists, we have learned that the tragedy of the commons is not just confined to developing countries. Landscapes in Russia, eastern Europe, and central Asia are blighted by the ravages of industrialization without the protections insisted upon by countries like the United States. Forests have been destroyed by acid precipitation, huge lakes are nearly gone because of the use of their waters for irrigation, and public health has been damaged by pollution of the atmosphere and water supplies.

The maps and data sets in the *Student Atlas of Environmental Issues* can help you, as a student, understand the dimensions of the world's environmental problems and the geographical basis of these problems. The maps are not perfect representations of reality—no maps ever are—but they do represent "models," or approximations of the real world, that should aid in your understanding of global environmental changes. Use this atlas in conjunction with your text and it will help you become more knowledgeable of the vast and complex processes of human modification of environmental systems.

A Word about Data Sources

At the very outset of your study of this atlas, you should be aware of some limitations of the maps and data tables. In some instances, a map or a table may have missing data. This may be the result of the failure of a country to report information to a central international body (such as the United Nations or the World Bank). Alternatively, it may reflect shifts in political boundaries, internal conflicts, or changes in responsibility for reporting data that have caused certain countries to delay their reports. It is always our wish to be as up-to-date as possible; subsequent editions of this atlas will probably have increased data. In the meantime, as events continue to restructure our world, it's an exciting time to be a student of environmental problems!

You will find your study of this atlas more productive in relation to your study of environmental issues if you examine the maps on the following pages in terms of five distinct analytical themes:

1. *Location: Where Is It?* This involves a focus on the precise location of places in both *absolute* terms (the latitude and longitude of a place) and *relative* terms (the location of a place in relation to the location of other places). When you think of location, you should automatically think of both forms. Knowing something about absolute location will help you to understand a variety of features of physical geography (such as climate, soil, and vegetation), since

such key elements are so closely related to their position on the earth. However, where places are located in relation to other places is often more important as a determinant of social, economic, and cultural characteristics that influence human/environment interaction than are the factors of physical geography.

2. *Place: What Is It Like?* This encompasses the economic, political, cultural, environmental, and other characteristics that give a place its identity. You should seek to understand the similarities and differences between places by exploring their basic characteristics. Why are some places with similar environmental characteristics so very different in economic, cultural, social, and political ways? Why are other places with such different environmental characteristics so seemingly alike in terms of their institutions, their economies, and their cultures?

3. *Human/Environment Interactions: How Is the Landscape Shaped?* This focuses on the ways in which people respond to and modify their environments. Throughout the world, the characteristics of the environment do not exert a controlling influence over human activities; they only provide a set of alternatives from which different cultures, in different times, make their choices. Observe the relationship between the basic elements of physical geography such as climate and terrain and the host of ways in which humans have used the land surfaces of the world.

4. *Movement: How Do People Stay in Touch?* This examines the transportation and communication systems that link people and places. Movement or "spatial interaction" is the chief mechanism for the spread of ideas and innovations from one place to another, validating the old cliché, "the world is getting smaller." Advanced transportation and communication systems have transformed the world into which your parents were born. And the world that greets your children will be very different from your world. It is largely because of the scale of spatial interaction that people in the United States can have environmental impacts on faraway places. The cheeseburger that you purchase at a fast-food chain store in Chicago may be made from beef raised in Middle American pastures that were created by destroying tropical rain forests.

5. *Regions: Worlds within a World.* This theme helps to organize knowledge about the land and its people. The world consists of a mosaic of "regions" or areas that are somehow distinctive and different from other areas. The regions of Anglo-America (the United States and Canada) and western Europe are classified by geographers as two clearly separate and environmentally distinct areas. Yet the two areas tend to be quite similar in terms of human/environmental interactions. Conversely, the regions of Anglo-America and East Asia have a number of shared physical environmental characteristics. But different cultural traditions and different institutions mean that human/environmental relationships in East Asia are going to be very different from those in Anglo-America. An understanding of both the differences and the similarities between such regions will help you to understand much about the ways in which people modify the environments that they occupy.

Not all of these concepts will be immediately apparent on each of the 48 maps and 14 tables in this atlas. But if you study the contents of *Student Atlas of Environmental Issues* as you read your text and you think about the five themes, the maps and tables and the text will complement one another and improve your understanding of global environmental issues.

Table of Contents

Part IV. Human Impact on Fresh Water and the Oceans   57

Part V. Human Impact on the Biosphere   78

Part VI. Human Impact on the Land   107

Part I

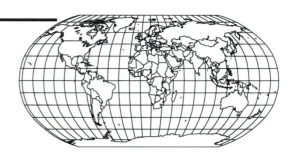

The Physical Environment: Global Patterns

Map 1 Current World Political Boundaries

Although environmental issues should, in theory, have no political base, the most important human components of the international environmental system are still individual countries or "states." The most crucial decisions impacting the global environment are made at the national level. The boundaries of countries are often, therefore, not only the most important source of political division in the world but the most important dividing lines of environmental policy as well.

Scale: 1 to 125,000,000

Note: All world maps are Robinson projection.

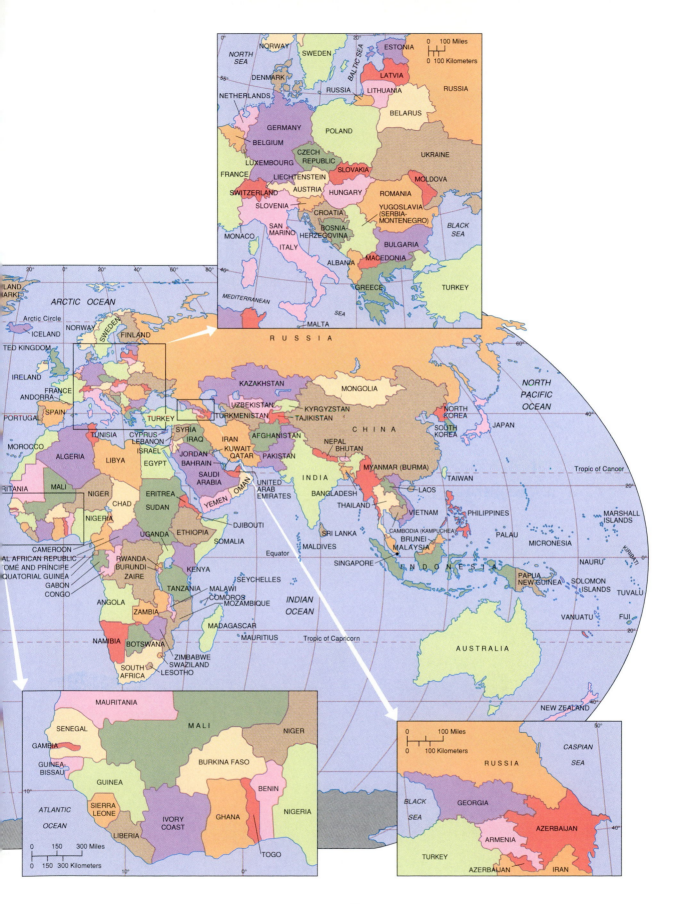

Map 2 World Climate Regions

Climates of the World

Tropical Moist Climates
- Rain Forest
- Monsoon
- Savanna

Dry Climates
- Warm Desert
- Warm Steppe
- Cool Desert
- Cool Steppe

Midlatitude Climates
- Summer Dry or Mediterranean
- Moist Subtropical
- Marine West Coast
- Cool Forest
- Subarctic

Polar Climates
- Tundra
- Ice Cap

Azonal Climates
- Undifferentiated Highlands

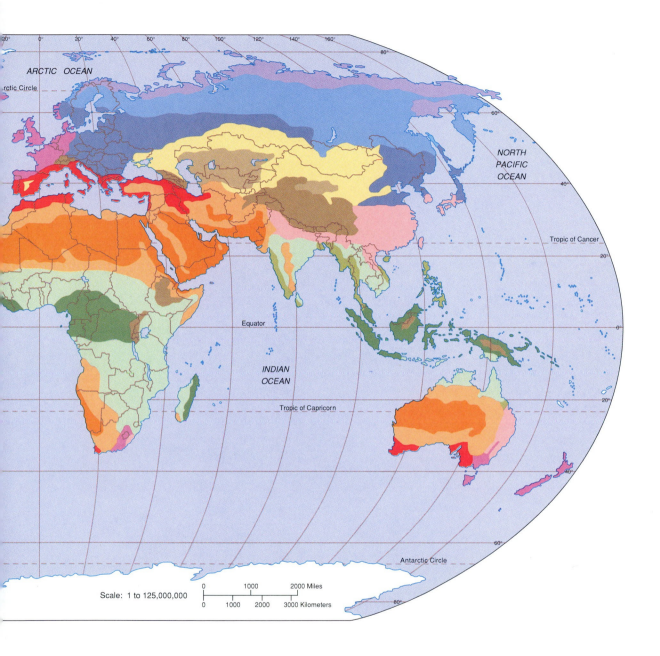

Of the world's many physical geographic features, climate (the long-term average of such weather conditions as temperature and precipitation) is the most important. It is climate that conditions the types of natural vegetation patterns and the types of soil that will exist in an area. It is also climate that determines the availability of our most precious resource: water. From an economic standpoint, the world's most important activity is agriculture; no other element of physical geography is more important for agriculture than climate. And of all the human activities that exert influence over the global environmental systems, none is more important than agriculture.

Map 3 Average Annual Precipitation

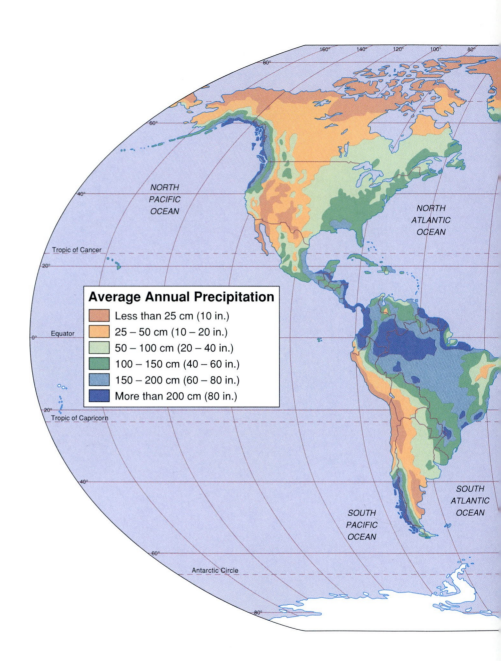

Average Annual Precipitation

- Less than 25 cm (10 in.)
- 25 – 50 cm (10 – 20 in.)
- 50 – 100 cm (20 – 40 in.)
- 100 – 150 cm (40 – 60 in.)
- 150 – 200 cm (60 – 80 in.)
- More than 200 cm (80 in.)

NORTH PACIFIC OCEAN

NORTH ATLANTIC OCEAN

SOUTH PACIFIC OCEAN

SOUTH ATLANTIC OCEAN

Tropic of Cancer

Equator

Tropic of Capricorn

Antarctic Circle

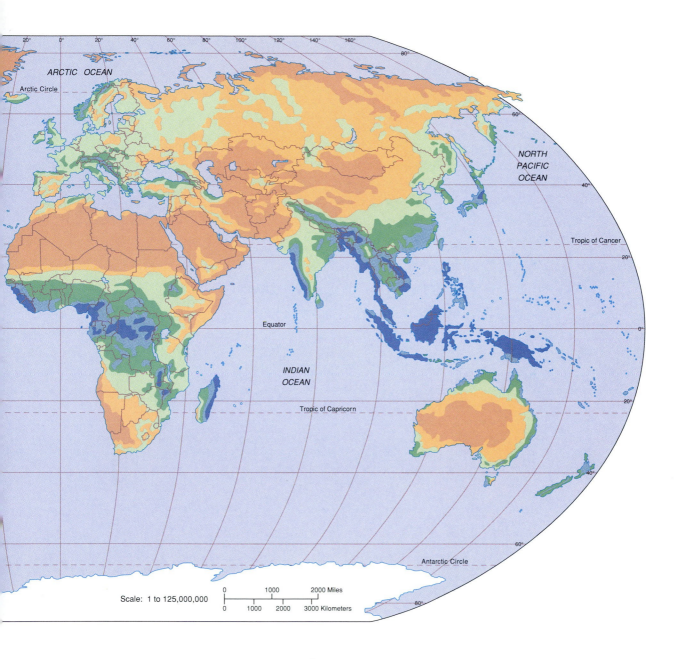

The two most important environmental variables are precipitation and temperature, the essential elements of weather and climate. Precipitation is a conditioner of both soil type and vegetation. More than any other single environmental element, it determines where people live and do not live. Water is the most precious resource available to humans, and water availability is largely a function of precipitation.

Map 4 Temperature Regions and Ocean Currents

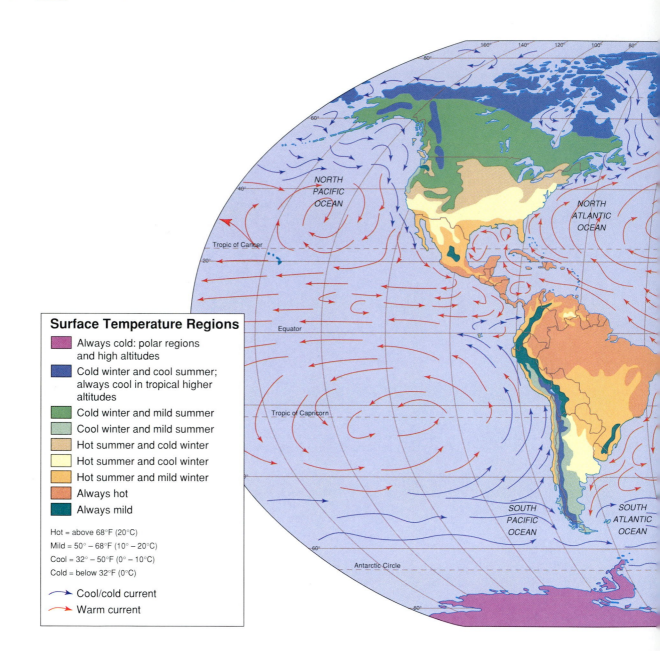

Surface Temperature Regions

- Always cold: polar regions and high altitudes
- Cold winter and cool summer; always cool in tropical higher altitudes
- Cold winter and mild summer
- Cool winter and mild summer
- Hot summer and cold winter
- Hot summer and cool winter
- Hot summer and mild winter
- Always hot
- Always mild

Hot = above 68°F (20°C)
Mild = 50° – 68°F (10° – 20°C)
Cool = 32° – 50°F (0° – 10°C)
Cold = below 32°F (0°C)

→ Cool/cold current
→ Warm current

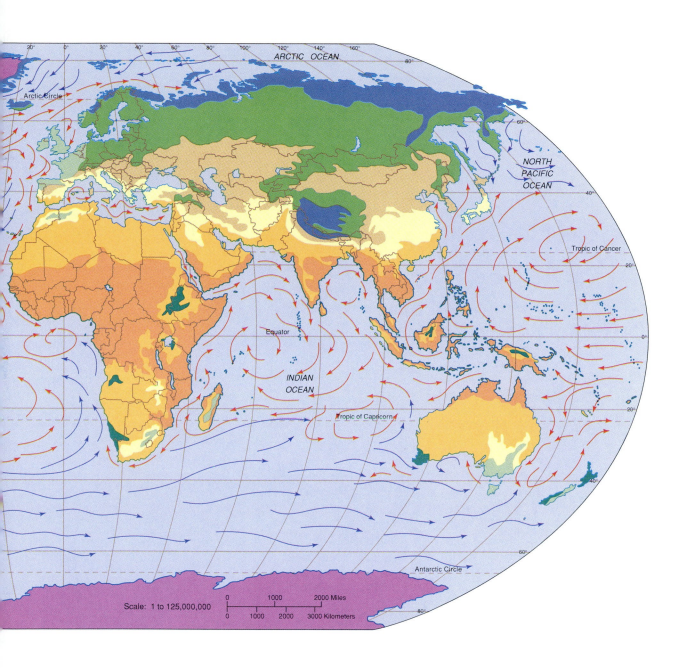

Along with precipitation, temperature is one of the two most important environmental variables, defining the climatic conditions so essential for the distribution of human activities and the human population. Ocean currents exert a significant influence over the climate of adjacent continents and are the most important mechanism for redistributing surplus heat from the equatorial region into middle and high latitudes.

Map 5 Vegetation

Vegetation Regions

- Needleleaf Forest
- Broadleaf Forest
- Mixed Forest (Broadleaf and Needleleaf)
- Woodland and Shrub (Mediterranean)
- Short Grass (Steppe)
- Tall Grass (Prairie)
- River Valley and Oasis
- Highlands (Unclassified; Vertical Zonation)
- Desert and Desert Shrub
- Savanna Grassland and Shrub
- Wooded Savanna
- Tropical Woodland and Shrub
- Light Tropical Forest
- Tropical Rain Forest
- Heath and Moor
- Tundra and Alpine Vegetation
- Permanent Ice Cover

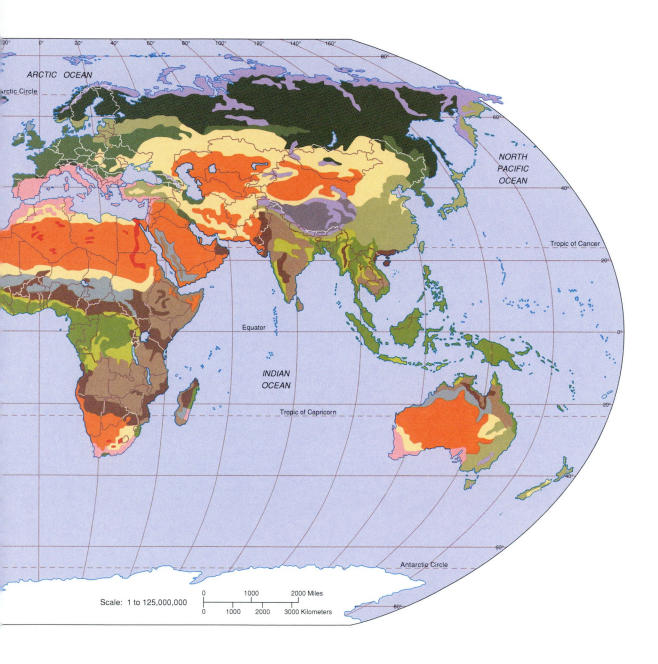

Vegetation is the most visible consequence of the distribution of temperature and precipitation. The global distribution of vegetation types and the global distribution of climate are closely related. But not all vegetation types are the consequence of temperature and precipitation or other climatic variables. Many types of vegetation, in many areas of the world, are the consequence of human activities, particularly the grazing of domesticated livestock, burning, and forest clearance.

Map 6 World Soil Orders

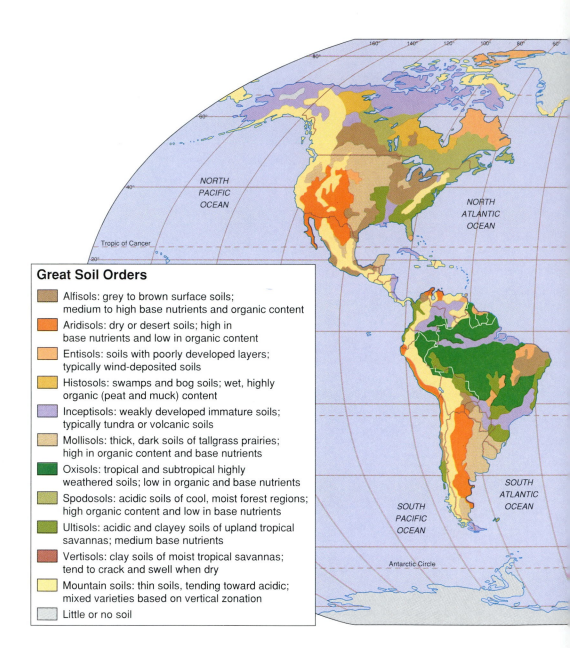

Great Soil Orders

Alfisols: grey to brown surface soils; medium to high base nutrients and organic content

Aridisols: dry or desert soils; high in base nutrients and low in organic content

Entisols: soils with poorly developed layers; typically wind-deposited soils

Histosols: swamps and bog soils; wet, highly organic (peat and muck) content

Inceptisols: weakly developed immature soils; typically tundra or volcanic soils

Mollisols: thick, dark soils of tallgrass prairies; high in organic content and base nutrients

Oxisols: tropical and subtropical highly weathered soils; low in organic and base nutrients

Spodosols: acidic soils of cool, moist forest regions; high organic content and low in base nutrients

Ultisols: acidic and clayey soils of upland tropical savannas; medium base nutrients

Vertisols: clay soils of moist tropical savannas; tend to crack and swell when dry

Mountain soils: thin soils, tending toward acidic; mixed varieties based on vertical zonation

Little or no soil

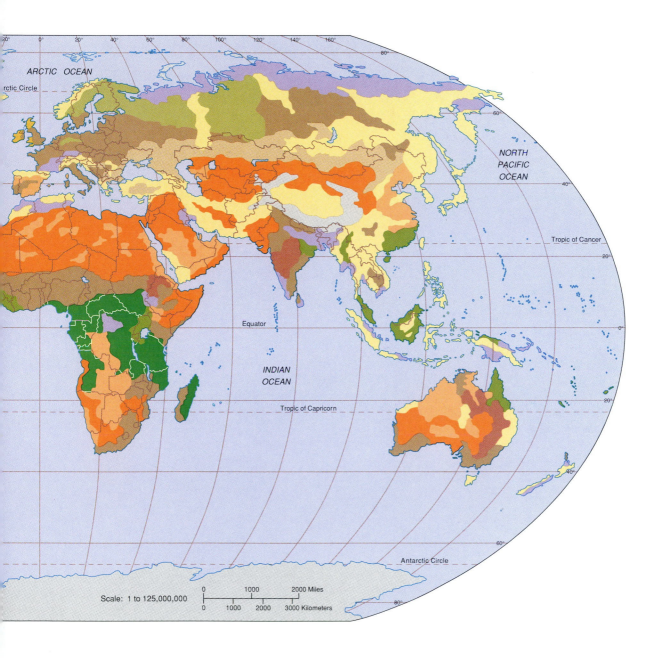

Soils are one of the three primary ecological factors, along with climate and vegetation, that determine the "habitability" of regions for humans. In particular, soils
influence the kinds of agricultural uses to which land is put. Since soils support the
plants that are the primary producers of all food in the terrestrial food chain, their
characteristics are crucial to the health and stability of ecosystems. Two types of
soil are shown on this map: zonal soils, whose characteristics are based on climate;
and azonal soils, such as alluvial (water-deposited) or aeolian (wind-deposited) soils,
whose characteristics are derived from forces other than climate.

Map 7 World Ecological Regions

World Ecological Regions

Arctic and Subarctic Zone

Ice Cap

Tundra Province: moss-grass and moss-lichen tundra

Tundra Altitudinal Zone: polar desert (no vegetation)

Subarctic Province: evergreen forest, needleleaf taiga; mixed coniferous and small-leafed forest

Subarctic Altitudinal Zone: open woodland; wooded tundra

Humid Temperate Zone

Moderate Continental Province: mixed coniferous and broadleaf forest

Moderate Continental Altitudinal Zone: coastal and alpine forest; open woodland

Warm Continental Province: broadleaf deciduous forest

Warm Continental Altitudinal Zone: upland broadleaf and alpine needleleaf forest

Marine Province: lowland, west-coastal humid forest

Marine Altitudinal Zone: humid coastal and alpine coniferous forest

Humid Subtropical Province: broadleaf evergreen and broadleaf deciduous forest

Humid Subtropical Altitudinal Zone: upland, subtropical broadleaf forest

Prairie Province: tallgrass and mixed prairie

Prairie Altitudinal Zone: upland mixed prairie and woodland

Mediterranean Province: sclerophyll woodland, shrub, and steppe grass

Mediterranean Altitudinal Zone: upland shrub and steppe

Humid Tropical Zone

Savanna Province: seasonally dry forest; open woodland; tallgrass savanna

Savanna Altitudinal Zone: open woodland steppe

Rain Forest Province: constantly humid, broadleaf evergreen forest

Rain Forest Altitudinal Zone: broadleaf evergreen and subtropical deciduous forest

Arid and Semiarid Zone

Tropical/Subtropical Steppe Province: dry steppe (short grass), desert shrub, semidesert savanna

Tropical/Subtropical Steppe Altitudinal Zone: upland steppe (short grass) and desert shrub

Tropical/Subtropical Desert Province: hot, lowland desert in subtropical and coastal locations; xerophytic vegetation

Tropical/Subtropical Desert Altitudinal Zone: desert shrub

Temperate Steppe Province: medium to shortgrass prairie

Temperate Steppe Altitudinal Zone: alpine meadow and coniferous woodland

Temperate Desert Province: midlatitude rainshadow desert; desert shrub

Temperate Desert Altitudinal Zone: extreme continental desert steppe; desert shrub, xerophytic vegetation, shortgrass steppe

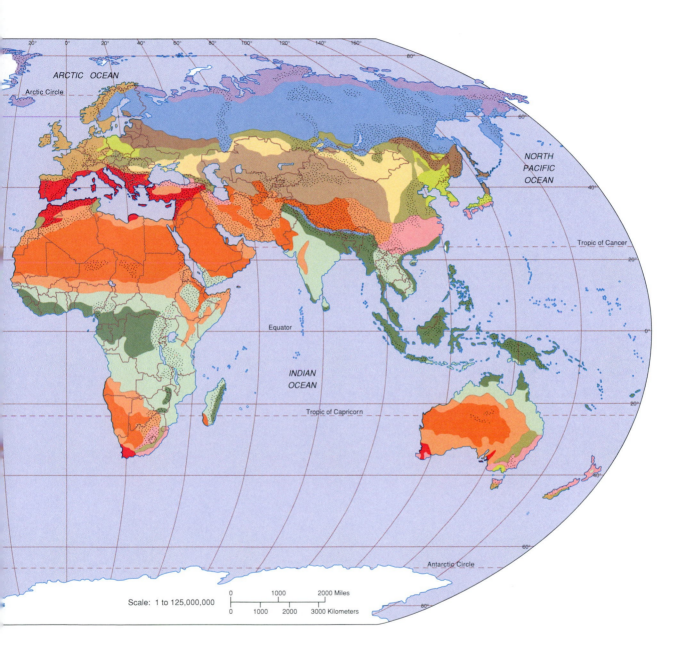

Scale: 1 to 125,000,000

Ecology is the study of the relationships between living organisms and their environmental surroundings. Ecological regions are distinctive areas within which unique sets of organisms and environments are found. Within each ecological region, a particular combination of vegetation, wildlife, soil, water, climate, and terrain defines that region's "habitability," or ability to support life, including human life. Like climate and landforms, ecological relationships are crucial to the existence of agriculture, the most basic of the economic relationships between humans and environmental systems.

Map 8 World Natural Hazards

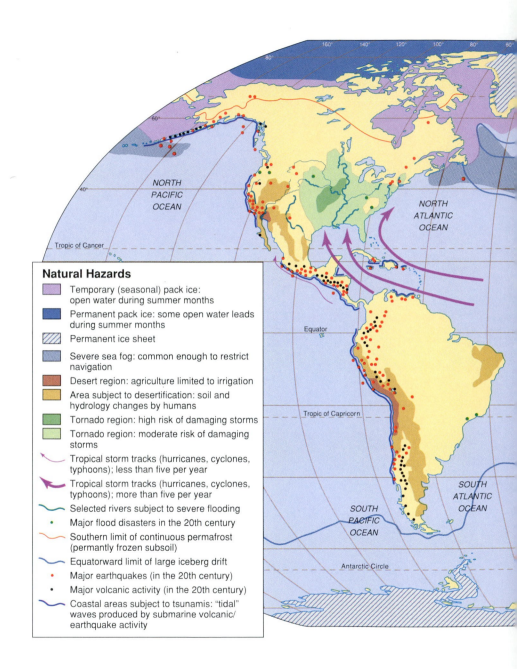

Natural Hazards

- Temporary (seasonal) pack ice: open water during summer months
- Permanent pack ice: some open water leads during summer months
- Permanent ice sheet
- Severe sea fog: common enough to restrict navigation
- Desert region: agriculture limited to irrigation
- Area subject to desertification: soil and hydrology changes by humans
- Tornado region: high risk of damaging storms
- Tornado region: moderate risk of damaging storms
- Tropical storm tracks (hurricanes, cyclones, typhoons); less than five per year
- Tropical storm tracks (hurricanes, cyclones, typhoons); more than five per year
- Selected rivers subject to severe flooding
- Major flood disasters in the 20th century
- Southern limit of continuous permafrost (permanently frozen subsoil)
- Equatorward limit of large iceberg drift
- Major earthquakes (in the 20th century)
- Major volcanic activity (in the 20th century)
- Coastal areas subject to tsunamis: "tidal" waves produced by submarine volcanic/earthquake activity

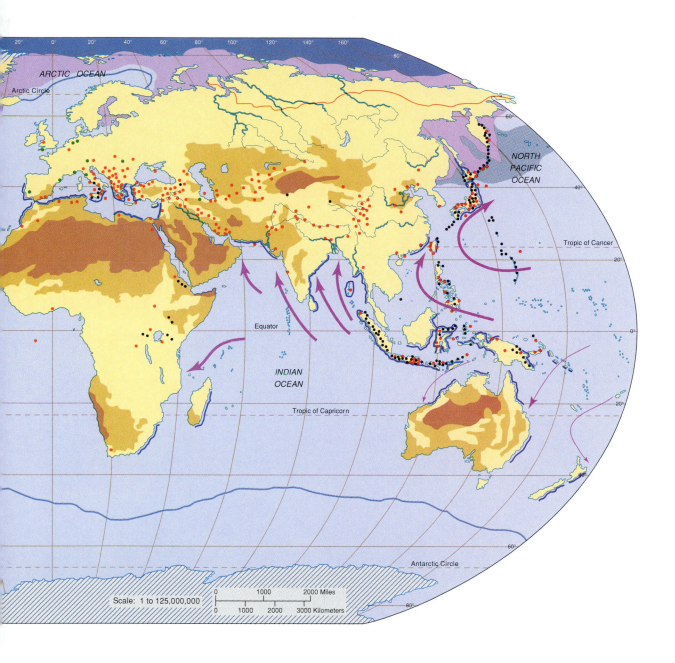

Unlike other elements of physical geography, natural hazards are unpredictable. However, there are certain regions where the *probability* of the occurrence of a particular natural hazard is high. This map shows regions affected by major natural hazards at rates that are higher than the global norm. The presence of persistent natural hazards may condition the types of modifications that people make in environments.

Map 9 World Topography

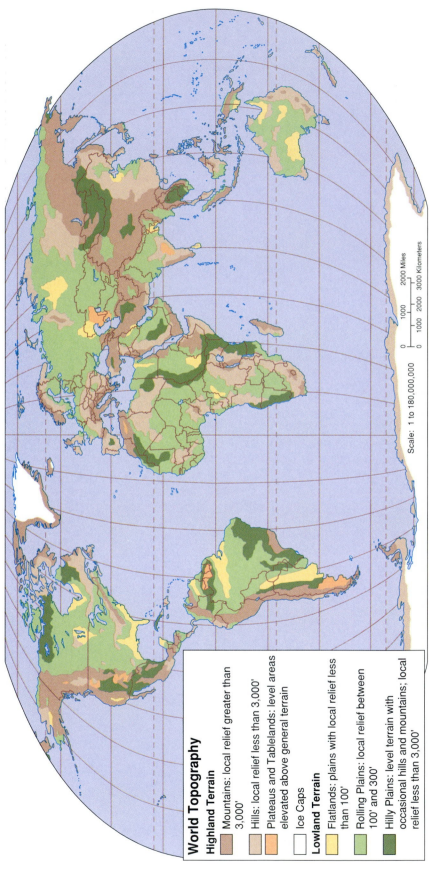

World Topography

Highland Terrain

- Mountains: local relief greater than 3,000'
- Hills: local relief less than 3,000'
- Plateaus and Tablelands: level areas elevated above general terrain
- Ice Caps

Lowland Terrain

- Flatlands: plains with local relief less than 100'
- Rolling Plains: local relief between 100' and 300'
- Hilly Plains: level terrain with occasional hills and mountains; local relief less than 3,000'

Scale: 1 to 180,000,000

```
0          1000       2000 Miles
0   1000   2000   3000 Kilometers
```

Second only to climate as a conditioner of human activity—particularly in agriculture and also in the location of cities and industry—is topography or terrain. It is what we often call "landforms." A comparison of this map with the map of land use (Map 11) will show that most of the world's productive agricultural zones are located in the lowland regions. Where large regions of agricultural productivity are found, we also tend to find urban concentrations and, with cities, we find industry. There is also a good spatial correlation between the map of landforms and the map showing the distribution and density of the human population (Map 10). The landforms shown on this map are primarily the result of extremely gradual geologic activity, such as the long-term movement of crustal plates (sometimes called "continental drift"). This activity occurs over hundreds of millions of years. Also important are the more rapid (but still slow by human standards) erosional activity of water, wind, glacial ice, and waves, tides, and currents. Significant erosional activity occurs over the span of a few million years. Some landforms may be produced by abrupt or "cataclysmic" events such as a major volcanic eruption or a meteor strike, but these events are relatively rare and their effects are usually too minor to show up on a map of this scale.

Part II

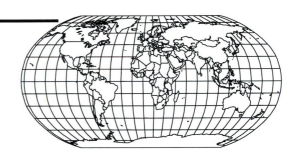

The Human Environment: Global Patterns

Map 10 World Population Density

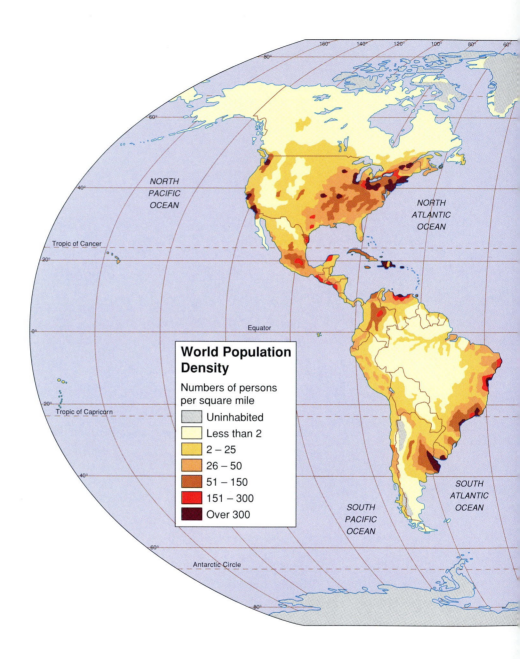

World Population Density

Numbers of persons per square mile

- Uninhabited
- Less than 2
- 2 – 25
- 26 – 50
- 51 – 150
- 151 – 300
- Over 300

NORTH PACIFIC OCEAN

NORTH ATLANTIC OCEAN

Tropic of Cancer

Equator

Tropic of Capricorn

SOUTH PACIFIC OCEAN

SOUTH ATLANTIC OCEAN

Antarctic Circle

No feature of human activity is more reflective of environmental conditions than where people live. In the areas of densest population, a mixture of natural and human factors have combined to allow maximum food production, maximum urbanization, and especially concentrated economic activity. Three such great concentrations appear on the map—East Asia, South Asia, and Europe—with a fourth lesser concentration in eastern North America (the "Megalopolis" region of the United States and Canada). One of these great population clusters—South Asia—is still growing rapidly and can be expected to become even more densely populated by the beginning of the twenty-first century. The other concentrations are likely to remain about as they are now. In Europe and North America, this is the result of

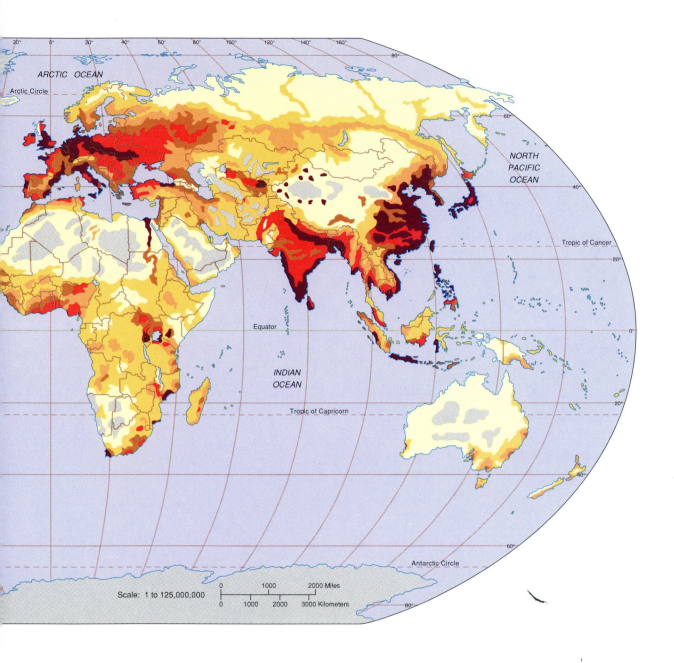

economic development that has caused population growth to level off during the last century. In East Asia, population has also begun to grow more slowly. In the case of Japan and the Koreas, this is the consequence of economic development; in the case of China, it is the consequence of government intervention in the form of strict family planning. The areas of future high density (in addition to those already existing) are likely to be in Middle and South America and Africa, where population growth rates are well above the world average. Population that is extremely dense or growing at an excessive rate when measured against a region's habitability is one of the greatest indicators of environmental deterioration.

Map 11 World Land Use

World Land Use

Predominant Activities by Region

- Manufacturing and Commerce
- Commercial Crop and Livestock Agriculture
- Intensive Subsistence Crop and Livestock Agriculture, including Plantations
- Tropical Shifting Subsistence Agriculture
- Livestock Ranching
- Dryland Nomadic Livestock Herding
- Forestry, Fishing, Hunting and Gathering, Recreation and Tourism (Commercial)
- Nomadic Herding, Forestry, Fishing, Hunting (Primarily Subsistence)
- Fishing Grounds (Commercial and Subsistence)
- No Major Economic Activity

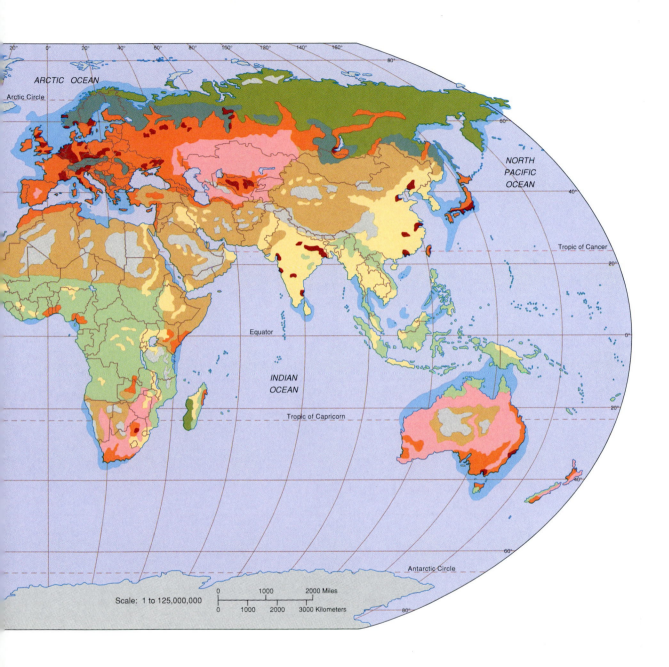

Many of the major land use patterns of the world (such as urbanization, industry, and transportation) are relatively small in area and are not easily seen on maps, but the most important uses people make of the earth's surface have more far-reaching effects. This map illustrates, in particular, the variations in primary land uses (such as agriculture) for the entire world. Note particularly the differences between land use patterns in the more developed countries of the temperate zones and the lesser developed countries of the tropics. These differences are often reflected in the type (although not necessarily in the extent) of environmental modification.

Map 12 World Urbanization

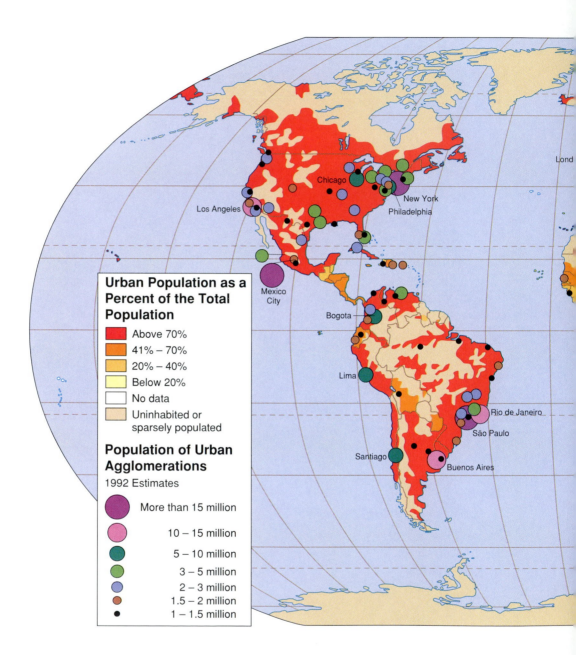

Urban Population as a Percent of the Total Population

- Above 70%
- 41% – 70%
- 20% – 40%
- Below 20%
- No data
- Uninhabited or sparsely populated

Population of Urban Agglomerations

1992 Estimates

- More than 15 million
- 10 – 15 million
- 5 – 10 million
- 3 – 5 million
- 2 – 3 million
- 1.5 – 2 million
- 1 – 1.5 million

Chicago
Los Angeles
New York
Philadelphia
Mexico City
Bogota
Lima
Rio de Janeiro
São Paulo
Santiago
Buenos Aires
Lond

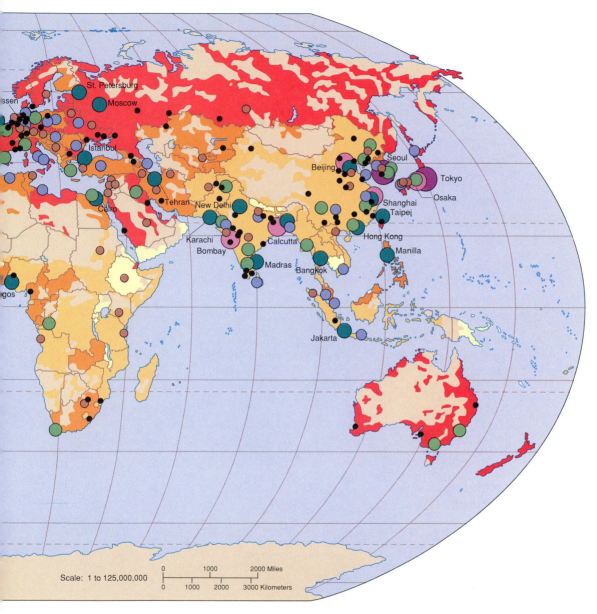

Scale: 1 to 125,000,000

A major indicator of the potential for environmental impact of humans in a particular region is the degree to which the region's population is concentrated in urban areas. Urban dwellers are rapidly becoming the norm among the world's people and rates of urbanization are increasing worldwide, with the greatest increases in urbanization taking place in developing regions. Whether in developing or developed countries, those who live in cities exert an influence on their environment that goes far beyond the confines of the city itself. Acting as the focal points for the flow of goods, cities draw resources not just from their immediate hinterlands but from the entire world, creating far-reaching environmental impacts as these resources are extracted, converted through industrial processes, and transported over great distances to metropolitan regions.

Map 13 Mineral Fuels and Critical Metals

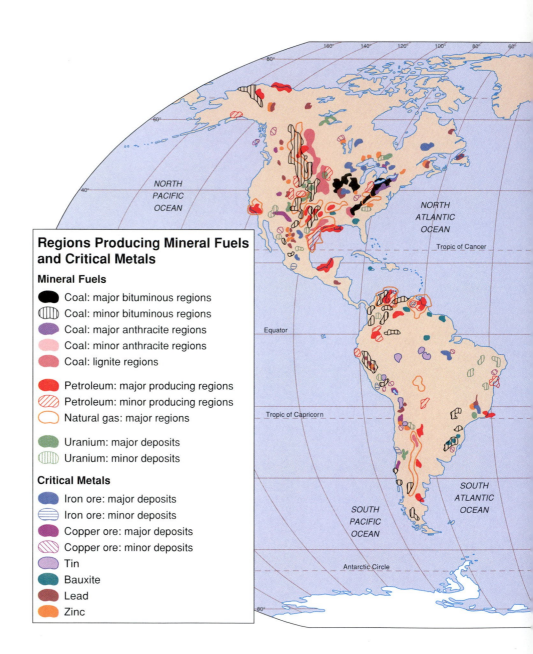

Regions Producing Mineral Fuels and Critical Metals

Mineral Fuels

- Coal: major bituminous regions
- Coal: minor bituminous regions
- Coal: major anthracite regions
- Coal: minor anthracite regions
- Coal: lignite regions

- Petroleum: major producing regions
- Petroleum: minor producing regions
- Natural gas: major regions

- Uranium: major deposits
- Uranium: minor deposits

Critical Metals

- Iron ore: major deposits
- Iron ore: minor deposits
- Copper ore: major deposits
- Copper ore: minor deposits
- Tin
- Bauxite
- Lead
- Zinc

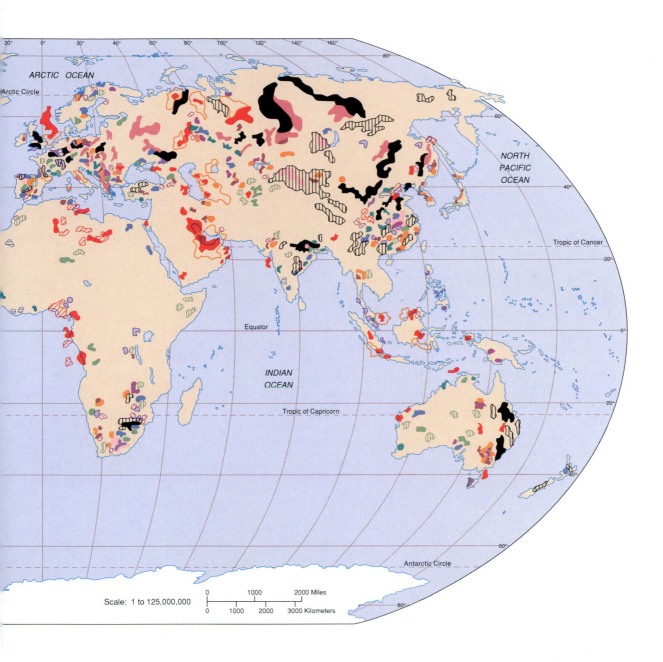

The extraction and transportation of mineral fuels (petroleum, natural gas, coal, and uranium) and of critical metals (copper and iron) produce environmental impacts second only to those of agriculture on a global scale. Nearly all of the most highly publicized environmental disasters of recent decades—including the Prince William Sound oil spill, the Chernobyl nuclear accident, and the Persian Gulf conflict oil spills—have involved mineral fuels that were being stored, transported, or used. And the continuing extraction of critical minerals produces high levels of atmospheric, soil, and water pollution. The location of fuels and metals tells us a great deal about where environmental degradation is likely to be occurring or to occur in the future.

Map 14 Surface Transportation Patterns

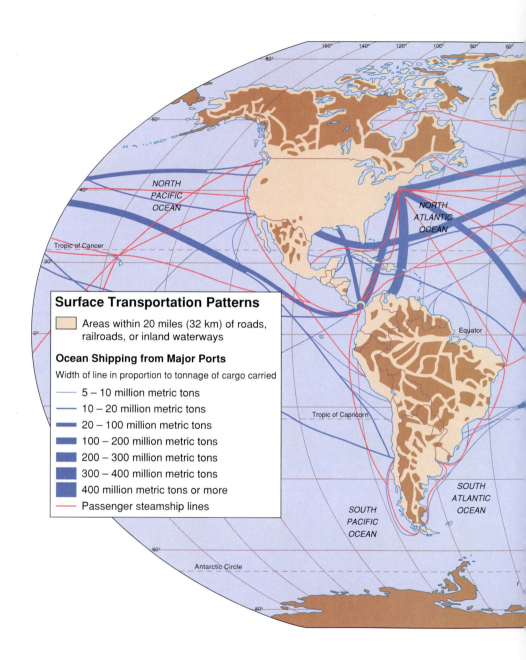

Surface Transportation Patterns

Areas within 20 miles (32 km) of roads, railroads, or inland waterways

Ocean Shipping from Major Ports

Width of line in proportion to tonnage of cargo carried

— 5 – 10 million metric tons

— 10 – 20 million metric tons

— 20 – 100 million metric tons

— 100 – 200 million metric tons

— 200 – 300 million metric tons

— 300 – 400 million metric tons

— 400 million metric tons or more

— Passenger steamship lines

NORTH PACIFIC OCEAN

NORTH ATLANTIC OCEAN

SOUTH PACIFIC OCEAN

SOUTH ATLANTIC OCEAN

Tropic of Cancer

Equator

Tropic of Capricorn

Antarctic Circle

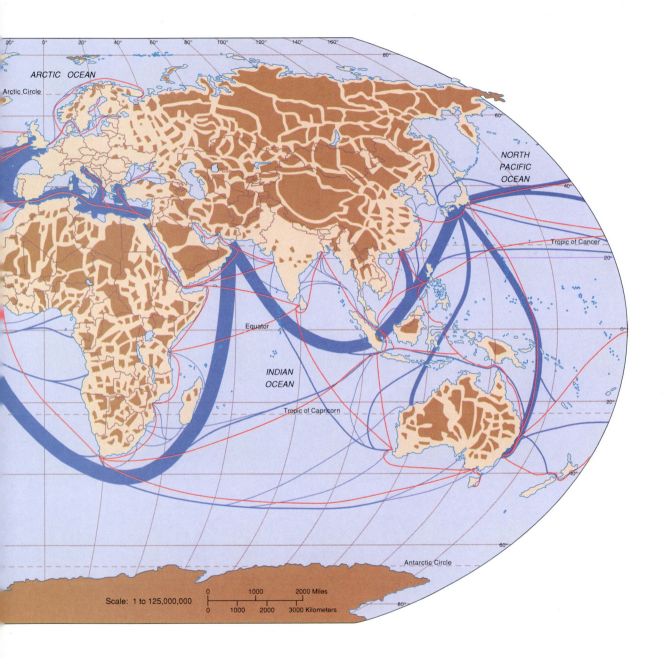

Scale: 1 to 125,000,000

0   1000   2000 Miles

0   1000   2000   3000 Kilometers

Transportation is second only to agriculture in its use of the earth's surface. Because transportation systems require significant modification of that surface, transportation is responsible for massive alterations in the quantity and quality of water, for major soil degradation and erosion, and (indirectly) for the air pollution that emanates from vehicles utilizing the transportation systems. In addition, as improved transportation technology draws together places on the earth that were formerly remote, it allows people to impact environments a great distance away from where they live.

Table A

World Countries: Area, Population, and Population Density, 1996

COUNTRY	AREA (mi^2)	POPULATION (1996 est.)[a]	DENSITY (pop/mi^2)[b]
Afghanistan	251,826	21,252,000	84
Albania	11,100	3,414,000	308
Algeria	919,595	28,539,000	31
Andorra	175	66,000	377
Angola	481,354	10,070,000	21
Antigua and Barbuda	171	65,000	380
Argentina	1,073,400	34,293,000	32
Armenia	11,506	3,557,000	309
Australia	2,966,155	18,322,000	6
Austria	32,377	7,987,000	247
Azerbaijan	33,436	7,790,000	233
Bahamas	5,382	257,000	48
Bahrain	267	576,000	2,157
Bangladesh	55,598	128,095,000	2,304
Barbados	166	256,000	1,542
Belarus	80,155	10,437,000	130
Belgium	11,783	10,082,000	856
Belize	8,866	214,000	24
Benin	43,475	5,523,000	130
Bhutan	18,200	1,781,000	98
Bolivia	424,165	7,896,000	19
Bosnia-Herzegovina	19,741	3,202,000	162
Botswana	231,800	1,392,000	6
Brazil	3,286,488	160,737,000	49
Brunei	2,226	292,000	131
Bulgaria	42,823	8,755,000	205
Burkina Faso	105,869	10,423,000	98
Burundi	10,745	6,262,000	582
Cambodia	69,898	10,561,000	151
Cameroon	183,569	13,521,000	74
Canada	3,849,674	28,435,000	7
Cape Verde	1,557	436,000	280
Central African Republic	240,535	3,210,000	14
Chad	495,755	5,587,000	11
Chile	292,135	14,161,000	48
China	3,689,631	1,203,097,000	326
Colombia	440,831	35,200,000	80
Comoros	863	549,000	636
Congo	132,047	2,505,000	19
Costa Rica	19,730	3,419,000	173
Croatia	21,829	4,666,000	214
Cuba	42,804	10,938,000	255
Cyprus	3,593	737,000	205
Czech Republic	30,613	10,433,000	341
Denmark	16,638	5,199,000	312
Djibouti	8,958	510,000	57
Dominica	305	83,000	272
Dominican Republic	18,704	7,948,000	425
Ecuador	109,484	10,891,000	99
Egypt	386,662	62,360,000	161
El Salvador	8,124	5,870,000	723
Equatorial Guinea	10,831	420,000	39
Eritrea	45,300	3,579,000	79
Estonia	17,413	1,625,000	93
Ethiopia	483,123	55,979,000	116
Fiji	7,078	773,000	109
Finland	130,559	5,085,000	39
France	211,208	58,109,000	275
Gabon	103,347	1,156,000	11
Gambia	4,127	989,000	239
Georgia	26,911	5,726,000	213
Germany	137,882	81,338,000	590
Ghana	92,098	17,763,000	193
Greece	50,962	10,648,000	209
Grenada	133	94,000	707

(Continued on next page)

COUNTRY	AREA (mi^2)	POPULATION (1996 est.)[a]	DENSITY (pop/mi^2)[b]
Guatemala	42,042	10,999,000	262
Guinea	94,926	6,549,000	69
Guinea-Bissau	13,948	1,125,000	81
Guyana	83,000	724,000	8
Haiti	10,714	6,518,000	608
Honduras	43,277	5,460,000	126
Hungary	35,920	10,319,000	287
Iceland	36,769	266,000	7
India	1,237,062	936,546,000	756
Indonesia	742,410	203,584,000	274
Iran	632,457	64,625,000	102
Iraq	169,235	20,644,000	122
Ireland	27,137	3,550,000	131
Israel[c]	8,019	5,143,000	641
Italy	116,234	58,262,000	501
Ivory Coast	124,518	14,791,000	118
Jamaica	4,244	2,574,000	606
Japan	145,870	125,506,000	860
Jordan	35,135	4,101,000	117
Kazakhstan	1,049,156	17,377,000	17
Kenya	224,961	28,817,000	128
Kiribati	280	79,000	282
Korea, North	46,540	23,487,000	505
Korea, South	38,230	45,554,000	1,191
Kuwait	6,880	1,817,000	264
Kyrgyzstan	76,641	4,770,000	62
Laos	91,429	4,837,000	53
Latvia	24,595	2,763,000	112
Lebanon	4,015	3,696,000	921
Lesotho	11,720	1,993,000	170
Liberia	38,250	3,073,000	80
Libya	679,362	5,248,000	8
Liechtenstein	62	31,000	500
Lithuania	25,174	3,876,000	154
Luxembourg	998	405,000	406
Macedonia	9,928	2,160,000	218
Madagascar	226,658	13,862,000	61
Malawi	45,747	9,808,000	214
Malaysia	129,251	19,724,000	152
Maldives	115	261,000	2,250
Mali	478,767	9,375,000	20
Malta	122	370,000	3,032
Marshall Islands	70	56,000	802
Mauritania	395,956	2,263,000	6
Mauritius	788	1,127,000	1,430
Mexico	756,066	93,986,000	124
Micronesia	271	123,000	454
Moldova	13,012	4,490,000	345
Mongolia	604,829	2,494,000	4
Morocco[d]	275,114	28,169,000	102
Mozambique	308,642	18,150,000	59
Myanmar	261,228	45,104,000	173
Namibia	317,818	1,652,000	5
Nepal	56,827	21,561,000	386
Netherlands	16,133	15,543,000	963
New Zealand	103,519	3,407,000	33
Nicaragua	50,054	4,206,000	84
Niger	489,191	9,280,000	19
Nigeria	356,669	101,232,000	284
Norway	125,050	4,331,000	35
Oman	82,030	2,125,000	26
Pakistan	310,432	131,542,000	424
Palau	188	17,000	88
Panama	29,157	2,681,000	92
Papua New Guinea	178,704	4,295,000	24
Paraguay	157,048	5,358,000	34
Peru	496,225	24,087,000	49
Philippines	115,831	73,087,000	633
Poland	120,728	38,792,000	321

(Continued on next page)

COUNTRY	AREA (mi^2)	POPULATION (1996 est.)[a]	DENSITY (pop/mi^2)[b]
Portugal	35,516	10,562,000	297
Qatar	4,416	534,000	121
Romania	91,699	23,198,000	253
Russia	6,592,849	149,909,000	23
Rwanda	10,169	8,605,000	846
St. Kitts and Nevis	104	41,000	394
St. Lucia	238	156,000	655
St. Vincent/Grenadines	150	118,000	787
San Marino	24	24,000	1,013
São Tomé and Príncipe	372	140,000	376
Saudi Arabia	830,000	18,730,000	23
Senegal	75,951	9,907,000	130
Seychelles	175	73,000	417
Sierra Leone	27,925	4,753,000	170
Singapore	246	2,889,000	11,744
Slovakia	18,933	5,432,000	287
Slovenia	7,819	2,052,000	262
Solomon Islands	10,954	399,000	36
Somalia	246,201	7,348,000	30
South Africa	433,680	45,095,000	104
Spain	194,885	39,404,000	202
Sri Lanka	24,962	18,343,000	735
Sudan	967,500	30,120,000	31
Suriname	63,251	430,000	7
Swaziland	6,704	967,000	144
Sweden	173,732	8,822,000	51
Switzerland	15,943	7,085,000	444
Syria	71,498	15,452,000	216
Taiwan	13,900	21,501,000	1,547
Tajikistan	55,251	6,155,000	111
Tanzania	364,900	28,701,000	79
Thailand	198,115	60,271,000	304
Togo	21,925	4,410,000	201
Tonga	290	106,000	366
Trinidad and Tobago	1,980	1,271,000	642
Tunisia	63,170	8,880,000	141
Turkey	300,948	63,406,000	211
Turkmenistan	188,456	4,075,000	22
Tuvalu	9	10,000	1,063
Uganda	93,104	19,573,000	210
Ukraine	233,090	51,868,000	223
United Arab Emirates	32,278	2,925,000	88
United Kingdom	94,248	58,295,000	618
United States	3,787,425	263,814,000	70
Uruguay	68,500	3,223,000	47
Uzbekistan	172,742	23,089,000	134
Vanuatu	4,707	174,000	37
Venezuela	352,145	21,005,000	60
Vietnam	128,066	74,393,000	581
Western Samoa	1,093	209,000	191
Yemen	205,356	14,728,000	72
Yugoslavia (Serbia-Montenegro)	39,449	11,102,000	281
Zaire	905,446	44,061,000	49
Zambia	290,586	9,446,000	33
Zimbabwe	150,873	11,140,000	74

[a]All populations are estimates for mid-1996, rounded to the nearest 1,000 of the population.

[b]Population densities as calculated by the UN. The figures are close to but not the same as may be derived by dividing the estimated 1996 population of a state by its area, as shown in this table.

[c]The figures for Israel do not include the West Bank and Gaza. The former has an area of approximately 2,300 square miles, a population of 1.7 million, and a density of 746 persons per square mile; the latter has an area of 100 square miles, a population of nearly 800,000 persons, and an estimated population density of 7,600 persons per square mile.

[d]Figures for Morocco include Western Sahara, annexed by Morocco.

Sources: U.S. Bureau of the Census, *World Population Profile;* United Nations Population Division; *Information Please Almanac 1996; The World Almanac and Book of Facts 1996.*

Table B
Agriculture and Food

	VALUE ADDED ($US millions)	CEREAL IMPORTS (thousands of tons)	FOOD AID IN CEREAL (thousands of tons)	FERTILIZER USE (hundreds of grams per hectare of arable land)	FOOD PRODUCTION PER CAPITA (annual % growth rate)
AFRICA					
Algeria	5,366	5,821	15	123	1.2
Angola	X[1]	X	X	X	X
Benin	760	134	19	82	1.9
Botswana	216	133	10	9	−2.1
Burkina Faso	X	121	30	60	2.5
Burundi	443	22	4	34	−0.3
Cameroon	3,170	281	1	30	−1.9
Central African Republic	584	32	5	5	−1.0
Chad	494	59	3	26	0.3
Congo	273	148	7	118	−1.5
Egypt	6,396	7,206	482	3,392	1.3
Equatorial Guinea	X	X	X	X	X
Eritrea	X	X	X	X	X
Ethiopia	3,476	X	X	95	−1.2
Gabon	447	77	X	11	−1.4
Gambia	83	87	6	44	−4.0
Ghana	2,893	396	75	38	0.3
Guinea	759	335	30	47	−0.3
Guinea-Bissau	108	70	9	10	0.9
Ivory Coast	3,026	590	41	132	−0.1
Kenya	1,357	569	287	410	−0.4
Lesotho	61	131	45	178	−2.2
Liberia	X	X	X	X	X
Libya	X	X	X	X	X
Madagascar	1,062	111	58	25	−1.5
Malawi	709	514	635	434	−4.2
Mali	1,128	83	34	103	−0.9
Mauritania	238	286	42	82	−1.6
Mauritius	274	240	5	2,512	0.0
Morocco	3,809	3,653	234	326	2.3
Mozambique	453	507	958	15	−2.1
Namibia	207	141	26	X	−2.0
Niger	855	136	26	4	−1.8
Nigeria	10,505	1,584	X	175	2.1
Rwanda	551	115	82	6	−2.5
Senegal	1,126	579	71	72	0.0
Sierra Leone	X	136	29	26	−1.2
Somalia	X	X	X	X	X
South Africa	4,815	2,275	X	596	−2.0
Sudan	X	X	X	X	X
Swaziland	X	X	X	X	X
Tanzania	1,168	215	35	137	−1.3
Togo	607	63	3	183	−0.6
Tunisia	2,287	1,044	100	223	1.5
Uganda	1,599	76	59	1	0.3
Zaire	X	X	X	X	X
Zambia	1,242	353	535	160	−0.3
Zimbabwe	757	538	900	481	−3.0
NORTH AND MIDDLE AMERICA					
Belize	X	X	X	X	X
Canada	X	1,095	X	479	0.6
Costa Rica	1,158	535	95	2,354	0.7
Cuba	X	X	X	X	X
Dominican Republic	1,473	961	7	694	−0.9
El Salvador	654	286	131	1,073	0.7
Guatemala	2,845	486	109	833	−0.5
Haiti	X	X	X	X	X
Honduras	566	197	64	210	−1.3

(Continued on next page)

	VALUE ADDED ($US millions)	CEREAL IMPORTS (thousands of tons)	FOOD AID IN CEREAL (thousands of tons)	FERTILIZER USE (hundreds of grams per hectare of arable land)	FOOD PRODUCTION PER CAPITA (annual % growth rate)
Jamaica	321	429	206	973	1.0
Mexico	29,037	6,223	45	653	−0.9
Nicaragua	545	125	85	246	−2.7
Panama	667	159	3	476	−1.2
Trinidad and Tobago	114	232	X	801	−0.6
United States	X	4,684	X	1,011	−0.3
SOUTH AMERICA					
Argentina	15,312	8	X	78	−0.3
Bolivia	X	298	227	58	0.7
Brazil	X	7,848	11	608	1.2
Chile	X	983	3	849	1.9
Colombia	7,607	1,702	17	1,032	0.7
Ecuador	1,746	428	14	380	0.6
Guyana	X	X	X	X	X
Paraguay	1,802	82	X	96	1.3
Peru	4,518	1,920	378	216	−0.4
Suriname	X	X	X	X	X
Uruguay	1,187	110	X	608	0.3
Venezuela	3,024	2,314	X	874	0.2
ASIA					
Afghanistan	X	X	X	X	X
Armenia	1,051	350	143	436	X
Azerbaijan	1,304	480	12	395	X
Bangladesh	7,306	1,175	719	1,032	−0.1
Bhutan	X	X	X	X	X
Cambodia	X	X	X	X	X
China	82,918	7,332	107	3,005	3.0
Georgia	1,738	500	170	680	X
India	70,702	694	276	720	1.5
Indonesia	27,189	3,105	40	1,147	2.2
Iran	25,653	4,840	31	755	1.0
Iraq	X	X	X	X	X
Israel	X	2,293	X	2,253	−1.8
Japan	80,528	28,035	X	3,951	−0.3
Jordan	353	1,596	254	398	0.2
Kazakhstan	X	100	3	134	X
Korea, North	X	X	X	X	X
Korea, South	23,403	22,271	X	4,656	0.5
Kuwait	110	251	X	1,600	X
Kyrgyzstan	X	120	91	242	X
Laos	685	8	8	42	−0.2
Lebanon	X	X	X	X	X
Malaysia	X	3,288	4	1,977	4.3
Mongolia	112	182	9	108	−2.5
Myanmar	X	X	X	69	−1.3
Nepal	1,532	27	15	391	1.2
Oman	374	369	X	1,270	X
Pakistan	11,500	2,893	188	1,015	1.2
Philippines	11,723	2,036	53	540	−1.3
Saudi Arabia	X	5,186	X	1,438	9.1
Singapore	X	X	X	X	X
Sri Lanka	2,311	1,149	248	964	−1.8
Syria	X	X	X	X	X
Tajikistan	X	450	72	1,618	X
Thailand	12,441	638	60	544	0.0
Turkey	23,609	2,107	2	702	0.3
Turkmenistan	X	940	2	1,204	X
United Arab Emirates	773	583	X	4,436	X
Uzbekistan	4,693	4,151	X	1,566	X
Vietnam	3,759	289	84	1,347	2.2
Yemen	2,511	1,843	21	99	−0.5

(Continued on next page)

	VALUE ADDED ($US millions)	CEREAL IMPORTS (thousands of tons)	FOOD AID IN CEREAL (thousands of tons)	FERTILIZER USE (hundreds of grams per hectare of arable land)	FOOD PRODUCTION PER CAPITA (annual % growth rate)
EUROPE					
Albania	277	647	513	338	−2.3
Austria	4,491	184	X	1,773	0.2
Belarus	4,643	1,250	246	2,228	X
Belgium	3,644	5,291	X	4,246	2.2
Bosnia-Herzegovina	X	X	X	X	X
Bulgaria	1,346	241	X	663	−1.9
Croatia	X	X	X	X	X
Czech Republic	1,952	519	X	X	X
Denmark	4,360	579	X	2,088	2.0
Estonia	411	46	231	1,229	X
Finland	4,717	108	X	1,363	−0.3
France	37,337	1,188	X	2,354	0.1
Germany	23,267	3,533	X	2,387	0.5
Greece	12,014	708	X	1,309	0.0
Hungary	2,135	137	X	292	−0.7
Iceland	X	X	X	X	X
Ireland	X	409	X	7,021	1.9
Italy	38,380	6,249	X	1,560	−0.3
Latvia	685	11	390	982	X
Lithuania	890	X	407	545	X
Macedonia	X	117	X	248	X
Moldova	1,485	200	72	612	X
Netherlands	11,636	4,431	X	5,889	0.4
Norway	X	302	X	2,276	0.2
Poland	5,434	3,142	200	811	0.7
Portugal	X	2,147	X	813	2.6
Romania	5,327	2,649	180	423	−2.4
Russia	35,553	11,238	1,124	417	X
Slovakia	741	X	X	X	X
Slovenia	583	549	X	2,306	X
Spain	20,295	4,955	X	769	1.1
Sweden	5,208	202	X	1,077	−1.4
Switzerland	X	455	X	3,340	−0.5
Ukraine	37,873	1,500	197	841	X
United Kingdom	16,383	3,534	X	3,205	0.0
Yugoslavia	X	X	X	X	X
OCEANIA					
Australia	9,404	32	X	265	0.3
Fiji	X	X	X	X	X
New Zealand	X	282	X	12,745	0.0
Papua New Guinea	1,321	227	0	308	−0.2
Solomon Islands	X	X	X	X	X

[1]X = Not Available

Source: World Development Report 1995.

Table C
World Countries: Energy Production, Consumption, and Requirements, 1993

	COMMERCIAL ENERGY				TRADITIONAL ENERGY		
	TOTAL PRODUCTION (in petajoules[1])	TOTAL CONSUMPTION (in petajoules)	PER CAPITA CONSUMPTION (in gigajoules)	IMPORTS AS % OF CONSUMPTION	TOTAL PRODUCTION (in petajoules)	PER CAPITA CONSUMPTION (in megajoules)	% OF TOTAL ENERGY CONSUMPTION
WORLD	**337,518**	**325,295**	**59**	**X**	**19,925**	**3,594**	**6**
AFRICA	**21,308**	**8,805**	**13**	**−134**	**4,815**	**6,991**	**35**
Algeria	4,587	1,183	44	−274	19	714	2
Angola	1,066	26	3	−3,835	56	5,455	58
Benin	13	7	1	−71	48	9,482	87
Botswana	X	X	X	X	13	9,420	100
Burkina Faso	X	8	1	100	85	8,652	91
Burundi	1	3	0	100	44	7,222	94
Cameroon	270	36	3	−639	114	9,130	75
Central African Republic	0	3	1	133	34	10,694	92
Chad	X	1	0	200	35	5,900	97
Congo	365	24	10	−1,379	22	8,945	48
Egypt	2,435	1,226	20	−84	45	752	4
Equatorial Guinea	0	2	5	100	4	11,522	69
Eritrea	X	X	X	X	X	X	X
Ethiopia	7	45	1	93	414	7,984	90
Gabon	637	32	26	−1,859	26	21,166	45
Gambia	X	3	3	100	9	8,579	75
Ghana	22	67	4	75	152	9,213	69
Guinea	1	15	2	100	35	5,594	70
Guinea-Bissau	X	3	3	100	4	4,012	58
Ivory Coast	18	109	8	119	103	7,723	49
Kenya	21	90	3	97	344	13,049	79
Lesotho	X	X	X	X	6	3,338	100
Liberia	1	5	2	100	48	17,045	91
Libya	3,054	457	91	−562	5	1,037	1
Madagascar	1	15	1	93	76	5,483	84
Malawi	3	11	1	82	133	12,596	92
Mali	1	7	1	100	54	5,279	88
Mauritania	0	39	18	118	0	37	0
Mauritius	0	21	19	143	17	15,392	44
Morocco	21	297	11	108	14	529	4
Mozambique	1	14	1	114	147	9,758	91
Namibia	X	X	X	X	X	X	X
Niger	5	15	2	67	47	5,484	76
Nigeria	4,140	705	7	−481	1,010	9,590	59
Rwanda	1	7	1	100	53	6,986	88
Senegal	X	38	5	126	49	6,257	57
Sierra Leone	X	6	1	233	30	6,903	83
Somalia	X	X	X	X	71	7,795	100
South Africa	4,146	3,578	79	−15	131	3,314	4
Sudan	3	48	2	110	220	8,261	82
Swaziland	X	X	X	X	18	22,852	100
Tanzania	2	30	1	100	330	11,769	92
Togo	0	9	2	100	10	2,265	53
Tunisia	209	218	25	7	31	3,593	12
Uganda	3	16	1	81	137	6,870	90
Zaire	78	73	2	4	365	8,854	83
Zambia	38	51	5	33	130	14,536	72
Zimbabwe	160	208	19	25	70	6,513	25
NORTH AND MIDDLE AMERICA	**87,427**	**97,154**	**220**	**11**	**1,825**	**4,130**	**2**
Belize	X	4	20	110	4	18,789	49
Canada	13,195	9,198	319	−43	67	2,325	1
Costa Rica	14	63	19	79	35	10,784	36

(Continued on next page)

| | COMMERCIAL ENERGY | | | | TRADITIONAL ENERGY | | |
	TOTAL PRODUCTION (in petajoules[1])	TOTAL CONSUMPTION (in petajoules)	PER CAPITA CONSUMPTION (in gigajoules)	IMPORTS AS % OF CONSUMPTION	TOTAL PRODUCTION (in petajoules)	PER CAPITA CONSUMPTION (in megajoules)	% OF TOTAL ENERGY CONSUMPTION
Cuba	43	369	34	95	205	18,848	36
Dominican Republic	6	148	20	96	25	3,360	15
El Salvador	21	72	13	74	39	7,050	35
Guatemala	22	72	7	89	104	10,335	59
Haiti	1	9	1	100	57	8,213	86
Honduras	8	43	8	81	58	10,897	57
Jamaica	0	104	43	100	5	2,493	5
Mexico	8,067	4,491	55	−57	248	2,755	5
Nicaragua	20	52	13	67	39	9,450	43
Panama	8	61	24	89	16	6,366	21
Trinidad and Tobago	470	267	209	−78	3	2,210	1
United States	65,547	81,751	317	21	916	3,553	1
SOUTH AMERICA	**15,355**	**10,095**	**33**	**−43**	**2,748**	**8,888**	**21**
Argentina	2,411	2,019	60	−9	116	3,421	5
Bolivia	164	86	12	−92	19	2,723	18
Brazil	2,491	3,800	24	46	2,021	12,912	35
Chile	222	539	39	62	84	6,050	13
Colombia	1,812	829	24	−117	235	6,927	22
Ecuador	784	245	22	−208	74	6,757	23
Guyana	9	15	18	100	4	5,355	23
Paraguay	113	51	11	−131	55	11,699	52
Peru	325	314	14	−3	88	3,825	22
Suriname	15	24	58	75	1	2,959	5
Uruguay	26	77	24	69	28	8,948	27
Venezuela	6,990	2,083	100	−226	22	1,046	1
ASIA	**113,332**	**95,679**	**28**	**−9**	**9,009**	**2,690**	**9**
Afghanistan	9	22	1	64	51	2,863	70
Armenia	11	49	14	96	0	X	X
Azerbaijan	714	546	74	23	0	0	0
Bangladesh	217	313	3	33	277	2,401	47
Bhutan	6	2	1	−150	12	7,345	85
Cambodia	0	7	1	100	54	5,560	88
China	31,359	29,679	25	−2	2,018	1,687	6
Georgia	28	159	29	91	X	X	X
India	8,088	9,338	10	21	2,824	3,132	23
Indonesia	7,145	2,648	14	−125	1,465	7,642	36
Iran	8,448	3,264	51	−164	29	446	1
Iraq	1,476	933	48	−24	1	53	0
Israel	1	505	96	118	0	24	0
Japan	3,466	17,505	141	87	10	78	0
Jordan	0	147	30	109	0	16	0
Kazakhstan	4,025	3,381	199	−16	0	0	0
Korea, North	2,671	2,925	127	8	40	1,753	1
Korea, South	832	4,504	102	98	26	584	1
Kuwait	4,329	471	265	−798	0	0	0
Kyrgyzstan	66	150	33	57	0	0	0
Laos	3	5	1	40	39	8,366	89
Lebanon	1	121	43	101	5	1,653	4
Malaysia	2,167	996	52	−114	90	4,686	8
Mongolia	84	105	45	19	13	5,689	11
Myanmar	75	71	2	3	193	4,324	73
Nepal	3	19	1	84	206	9,882	92
Oman	1,720	162	81	−957	0	0	0
Pakistan	766	1,135	9	36	296	2,228	21
Philippines	276	787	12	84	382	5,892	33
Saudi Arabia	19,171	2,933	171	0	0	0	0
Singapore	X	745	267	202	0	0	0
Sri Lanka	14	78	4	113	89	4,996	53
Syria	1,234	565	41	−105	0	9	0
Tajikistan	70	258	45	75	0	0	0

(Continued on next page)

	COMMERCIAL ENERGY				TRADITIONAL ENERGY		
	TOTAL PRODUCTION (in petajoules[1])	TOTAL CONSUMPTION (in petajoules)	PER CAPITA CONSUMPTION (in gigajoules)	IMPORTS AS % OF CONSUMPTION	TOTAL PRODUCTION (in petajoules)	PER CAPITA CONSUMPTION (in megajoules)	% OF TOTAL ENERGY CONSUMPTION
Thailand	678	1,628	28	63	526	9,141	24
Turkey	779	1,979	33	67	96	1,606	5
Turkmenistan	2,430	555	142	−327	X	X	X
United Arab Emirates	5,273	1,039	572	−364	0	0	0
Uzbekistan	X	1,903	87	−13	0	0	0
Vietnam	489	316	4	−48	251	3,516	44
Yemen	X	X	X	X	X	X	X
EUROPE	**92,937**	**108,523**	**148**	**18**	**552**	**761**	**1**
Albania	44	43	13	37	15	4,485	26
Austria	263	966	123	75	30	3,766	3
Belarus	122	1,249	123	91	X	X	X
Belgium	470	1,976	197	90	6	557	0
Bosnia-Herzegovina	14	29	8	52	X	X	X
Bulgaria	376	965	109	70	13	1,448	1
Croatia	179	263	58	43	0	0	0
Czech Republic	1,439	1,659	161	18	0	0	0
Denmark	525	762	148	31	5	943	1
Estonia	121	214	138	45	0	0	0
Finland	324	1,014	200	63	30	5,892	3
France	4,746	9,153	159	53	101	1,757	1
Germany	6,178	13,724	170	57	0	0	0
Greece	352	989	95	74	13	1,274	1
Hungary	533	990	97	53	24	2,319	2
Iceland	25	54	205	54	0	0	0
Ireland	152	428	121	67	0	139	0
Italy	1,226	6,749	118	74	48	848	1
Latvia	14	187	72	84	X	X	X
Lithuania	138	368	99	62	X	X	X
Macedonia	86	139	66	45	0	0	0
Moldova	1	234	53	102	X	X	X
Netherlands	3,112	3,306	216	13	2	150	0
Norway	6,365	904	210	−588	9	2,198	1
Poland	3,878	4,056	106	4	X	X	X
Portugal	35	603	61	106	6	573	1
Romania	1,345	1,762	7	30	19	841	1
Russia	43,550	30,042	203	−40	0	0	0
Slovakia	186	672	126	74	0	0	0
Slovenia	86	194	100	56	X	X	X
Spain	1,204	3,359	85	78	18	466	1
Sweden	950	1,660	191	46	122	14,062	7
Switzerland	386	985	139	57	14	2,052	1
Ukraine	4,501	8,058	156	46	0	0	0
United Kingdom	9,663	90,518	164	1	4	72	0
Yugoslavia	347	381	36	17	X	X	X
OCEANIA	**7,159**	**4,595**	**166**	**−61**	**185**	**6,693**	**4**
Australia	6,658	3,917	222	−77	109	6,191	3
Fiji	1	11	15	109	12	15,606	52
New Zealand	496	565	162	15	0	140	0
Papua New Guinea	2	33	8	97	60	14,550	64
Solomon Islands	X	2	6	X	3	9,107	62

[1]One petajoule = 1×10^{15} joules = 947,800,000,000 BTUs = 163,400 UN standard barrels of oil.
One gigajoule = 1×10^9 joules = 947,800 BTUs = .16 UN standard barrels of oil.
One megajoule = 1,000,000 joules = 947.8 BTUs = .00016 UN standard barrels of oil.

Sources: United Nations Statistical Division; *World Development Report 1995; World Resources 1996–97.*

Table D

Top Ten Producers and Consumers of Major Metals, 1985–1994

	PRODUCER 1985 (thousands of metric tons)	PRODUCER 1994 (thousands of metric tons)		CONSUMER 1985 (thousands of metric tons)	CONSUMER 1994 (thousands of metric tons)
ALUMINUM					
Australia	31,838.9	41,733.0	United States	4,282.0	5,407.1
Guinea	11,790.0	17,040.0	Japan	1,694.8	2,174.8
Jamaica	6,239.0	11,571.3	China	630.0	1,318.0
Brazil	5,846.0	8,280.8	Germany	1,390.9	1,300.0
China	1,650.0	7,260.0	Russia	1,750.0	1,185.0
India	2,281.0	5,280.0	France	586.1	665.0
Russia	4,600.0	4,000.0	South Korea	145.6	557.0
Suriname	3,738.0	3,200.5	Italy	470.0	554.0
Venezuela	0.0	2,540.0	United Kingdom	350.4	477.3
Greece	2,453.0	2,168.0	India	297.6	475.3
10 Country Total	**70,435.9**	**103,073.6**	**10 Country Total**	**11,597.4**	**14,113.5**
World Total	**84,189.0**	**111,024.2**	**World Total**	**14,861.5**	**20,201.1**
Bauxite, World Reserves[1]		**23,000,000**	**World Reserves Life Index, years**		**207**
CADMIUM					
Japan	2.5	2.6	Japan	1.9	6.6
Canada	1.7	2.2	Belgium	1.9	2.6
Belgium	1.3	1.6	United States	3.7	2.2
Russia	3.0	1.5	France	1.1	1.5
China	0.5	1.3	Russia	2.9	1.0
United States	1.6	1.1	United Kingdom	1.4	0.7
Germany	1.1	1.1	Germany	1.6	0.7
Australia	0.9	0.9	China	0.4	0.6
Italy	0.5	0.6	India	0.2	0.4
South Korea	0.1	0.6	South Korea	0.3	0.4
10 Country Total	**13.2**	**13.5**	**10 Country Total**	**13.4**	**16.7**
World Total	**19.1**	**18.3**	**World Total**	**17.6**	**18.3**
World Reserves		**540**	**World Reserves Life Index, years**		**X**
COPPER					
Chile	1,359.8	2,219.9	United States	1,958.0	2,674.3
United States	1,104.8	1,795.4	Japan	1,226.3	1,374.9
Canada	738.6	617.3	Germany	886.8	983.1
Russia	600.0	540.0	China	420.0	745.7
China	185.0	432.1	Russia	1,304.0	560.0
Australia	259.8	415.6	France	397.8	495.0
Zambia	452.6	384.4	South Korea	206.6	476.2
Poland	431.3	376.8	Italy	362.0	467.9
Peru	391.3	359.9	Belgium	309.6	404.9
Indonesia	88.7	333.8	United Kingdom	346.5	377.3
10 Country Total	**5,611.9**	**7,475.2**	**10 Country Total**	**7,418.6**	**8,559.3**
World Total	**8,088.2**	**9,522.6**	**World Total**	**9,699.9**	**11,084.2**
World Reserves		**310,000**	**World Reserves Life Index, years**		**33**

	PRODUCER 1985 (thousands of metric tons)	PRODUCER 1994 (thousands of metric tons)		CONSUMER 1985 (thousands of metric tons)	CONSUMER 1994 (thousands of metric tons)
LEAD					
Australia	498.0	523.8	United States	1,141.7	1,374.8
China	200.0	376.2	Germany	440.0	347.9
United States	424.4	374.0	Japan	394.9	345.0
Peru	201.5	216.7	United Kingdom	274.3	267.6
Canada	268.3	172.6	Italy	235.0	262.2
Mexico	206.7	164.4	France	208.0	246.7
Kazakhstan	440.0	160.0	China	220.0	214.1
Sweden	75.9	112.8	Russia	800.0	200.0
Nambia	34.5	93.1	South Korea	63.2	175.1
Morocco	106.8	75.7	Mexico	105.6	162.0
10 Country Total	**2,456.2**	**2,269.3**	**10 Country Total**	**3,882.7**	**3,595.4**
World Total	**3,431.2**	**2,764.7**	**World Total**	**5,236.7**	**5,342.2**
World Reserves		**63,000**	**World Reserves Life Index, years**		**23**
MERCURY					
China	0.7	0.7[2]	United States	1.7	1.2[2]
Algeria	0.8	0.4	Spain	0.6	0.8
Spain	0.9	0.3	Algeria	0.2	0.7
Kyrgyzstan	X	0.3	United Kingdom	0.3	0.4
Finland	0.1	0.1	China	0.4	0.3
United States	0.6	0.1	Brazil	0.2	0.3
Russia	2.2	0.1	Germany	0.3	0.2
Tajikistan	X	0.1	Mexico	0.2	0.2
Slovakia	0.2	0.1	Belgium	0.3	0.1
Ukraine	X	0.1	Russia	X	0.1
10 Country Total	**5.5**	**2.1**	**10 Country Total**	**4.1**	**4.2**
World Total	**6.8**	**2.9**	**World Total**	**7.4**	**6.6**
World Reserves		**130**	**World Reserves Life Index, years**		**45**
NICKEL					
Russia	185.1	243.0	Japan	136.1	164.9
Canada	170.0	150.1	United States	143.1	137.3
Indonesia	40.3	81.2	Germany	87.0	93.9
New Caledonia	72.4	73.6	Russia	138.0	64.0
Australia	85.8	71.9	Italy	29.0	44.6
Dominican Rep.	25.4	31.6	France	31.9	42.2
Cuba	32.1	31.0	United Kingdom	24.8	38.0
China	25.0	30.7	Finland	21.0	26.8
South Africa	25.0	30.1	China	14.7	23.4
Columbia	15.5	20.8	Sweden	17.0	23.0
10 Country Total	**676.6**	**764.0**	**10 Country Total**	**642.0**	**829.9**
World Total	**812.6**	**802.5**	**World Total**	**775.2**	**882.0**
World Reserves		**47,000**	**World Reserves Life Index, years**		**59**

	PRODUCER 1985 (thousands of metric tons)	PRODUCER 1994 (thousands of metric tons)		CONSUMER 1985 (thousands of metric tons)	CONSUMER 1994 (thousands of metric tons)
TIN					
China	15.0	46.0	United States	37.8	33.5
Indonesia	21.7	30.6	Japan	31.6	29.4
Peru	3.8	20.0	China	11.5	26.1
Brazil	26.5	17.0	Germany	17.8	18.2
Bolivia	16.1	16.1	Russia	31.5	14.5
Malaysia	36.9	6.5	United Kingdom	24.8	10.4
Australia	6.4	6.4	South Korea	2.6	9.8
Russia	13.5	5.0	France	6.9	9.2
Portugal	0.2	4.3	Netherlands	4.5	7.9
Thailand	16.9	3.1	Thailand	0.6	6.1
10 Country Total	**157.0**	**155.0**	**10 Country Total**	**169.6**	**164.1**
World Total	**180.7**	**169.4**	**World Total**	**215.4**	**216.8**
World Reserves		**7,000**	**World Reserves Life Index, years**		**41**
ZINC					
Canada	1,172.2	1,007.3	United States	962.0	1,118.3
Australia	759.1	945.0	Japan	780.0	723.1
China	300.0	900.0	China	349.0	611.9
Peru	523.4	602.6	Germany	480.0	531.6
United States	251.9	513.1	Italy	218.0	336.1
Mexico	275.4	369.7	Russia	1,000.0	330.0
Sweden	216.4	173.3	France	247.0	296.7
Kazkhstan	810.0	250.0	South Korea	120.0	264.9
North Korea	180.0	210.0	Belgium	169.0	225.0
Thailand	191.6	210.0	Australia	86.6	215.4
10 Country Total	**4,680.0**	**5,181.0**	**10 Country Total**	**4,411.6**	**4,553.0**
World Total	**6,125.0**	**6,895.1**	**World Total**	**6,552.0**	**6,960.3**
World Reserves		**140,000**	**World Reserves Life Index, years**		**20**
IRON ORE					
China	80,000.0	234,660.0	China	140,354.0	222,771.0
Brazil	128,251.0	151,000.0	Russia	203,760.0	168,938.0
Australia	97,447.0	120,534.0	Japan	102,215.0	113,783.0
Russia	247,639.0	75,000.0	United States	64,679.0	63,039.0
Ukraine	X	70,000.0	Brazil	36,419.0	44,985.0
India	42,545.0	61,000.0	Germany	45,204.0	41,350.0
United States	49,333.0	55,651.0	South Korea	11,709.0	32,001.0
Canada	39,502.0	30,668.0	France	26,606.0	20,199.0
South Africa	24,414.0	29,385.0	Belgium	13,353.0	17,975.0
Macedonia	X	20,000.0	United Kingdom	15,176.0	15,826.0
10 Country Total	**709,331.0**	**647,798.0**	**10 Country Total**	**659,475.0**	**740,847.0**
World Total	**860,640.0**	**988,797.0**	**World Total**	**860,640.0**	**970,422.0**
World Reserves		**150,000,000**	**World Reserves Life Index**		**152**

	PRODUCER 1985 (thousands of metric tons)	PRODUCER 1994 (thousands of metric tons)		CONSUMER 1985 (thousands of metric tons)	CONSUMER 1994 (thousands of metric tons)
STEEL, CRUDE[3]					
Japan	105,281.0	99,600.0	Russia	157,161.0	131,865.0
United States	80,069.0	88,793.0	Japan	73,377.0	99,149.0
China	46,721.0	88,680.0	United States	105,593.0	93,325.0
Russia	154,670.0	58,000.0	China	71,426.0	71,042.0
Germany	48,350.0	37,600.0	Germany	39,995.0	39,088.0
South Korea	132,539.0	33,000.0	Italy	21,680.0	26,593.0
Ukraine	X	30,500.0	South Korea	11,310.0	26,190.0
Italy	23,789.0	25,701.0	India	14,400.0	20,300.0
Brazil	20,456.0	25,000.0	France	14,812.0	16,588.0
India	12,185.0	16,500.0	United Kingdom	14,360.0	14,600.0
10 Country Total	**505,060**	**505,374.0**	**10 Country Total**	**524,306.0**	**536,740.0**
World Total	**718,131**	**725,129.0**	**World Total**	**720,568.0**	**732,202.0**

[1]Bauxite is the primary source of aluminum.

[2]Mercury figures in this column are for 1990.

[3]World reserves figures are not available for steel.

Source: World Resources 1996–97.

Part III

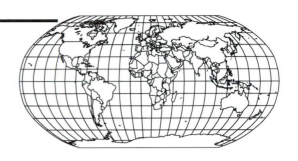

Human Impact on the Air

Map 15 Global Air Pollution: Sources and Wind Currents

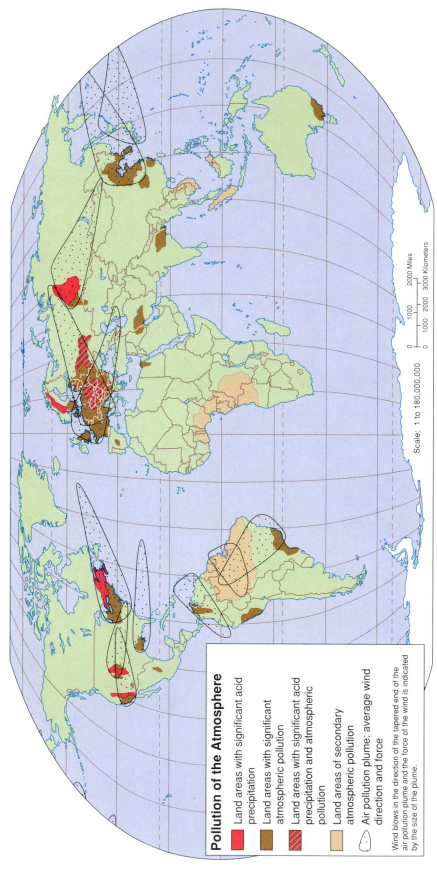

Scale: 1 to 180,000,000

0   1000   2000 Miles
0 1000 2000 3000 Kilometers

Pollution of the Atmosphere

Land areas with significant acid precipitation

Land areas with significant atmospheric pollution

Land areas with significant acid precipitation and atmospheric pollution

Land areas of secondary atmospheric pollution

Air pollution plume: average wind direction and force

Wind blows in the direction of the tapered end of the air pollution plume and the force of the wind is indicated by the size of the plume.

Almost all environmental processes begin and end with the flows of energy and matter among land, sea, and air. Because of the primacy of the atmosphere in this exchange system, air pollution is potentially one of the most dangerous human modifications in environmental systems. Pollutants such as nitrous and sulfur oxides cause the development of acid precipitation, which damages soil, vegetation, and wildlife and fish. Air pollution in the form of smog is often damaging to human health. And the efficiency of the atmosphere in holding heat—the so-called greenhouse effect—may be enhanced by increased CO_2, methane, and other gases produced by human agriculture and industry. The result could be a period of global warming that will dramatically alter climates in all parts of the world.

Map 16 The Acid Deposition Problem: Air, Water, Soil

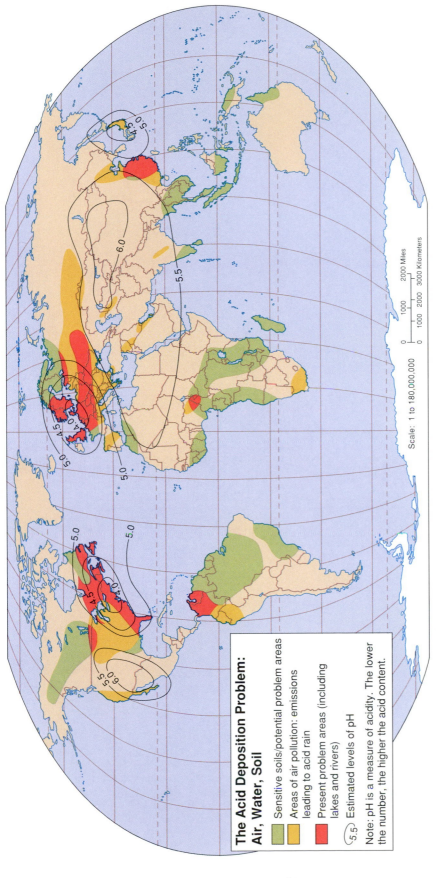

The Acid Deposition Problem: Air, Water, Soil

- 🟩 Sensitive soils/potential problem areas
- 🟨 Areas of air pollution: emissions leading to acid rain
- 🟥 Present problem areas (including lakes and rivers)
- 〰️5.5 Estimated levels of pH

Note: pH is a measure of acidity. The lower the number, the higher the acid content.

Scale: 1 to 180,000,000

0   1000   2000 Miles

0   1000   2000   3000 Kilometers

The term "acid precipitation" refers to increasing levels of acidity in snowfall and rainfall caused by atmospheric pollution. Oxides of nitrogen and oxygen resulting from incomplete combustion of fossil fuels (coal, oil, and natural gas) combine with water vapor in the atmosphere to produce weak acids that then "precipitate" or fall along with water or ice crystals. In some areas of the world, the increased acidity of streams and lakes stemming from high levels of acid precipitation has damaged aquatic life. Acid precipitation also harms soil systems and vegetation, producing a characteristic "burned" appearance in forests that lends the same quality to landscapes that forest fires would.

Map 17 Acid Precipitation: Eastern North America

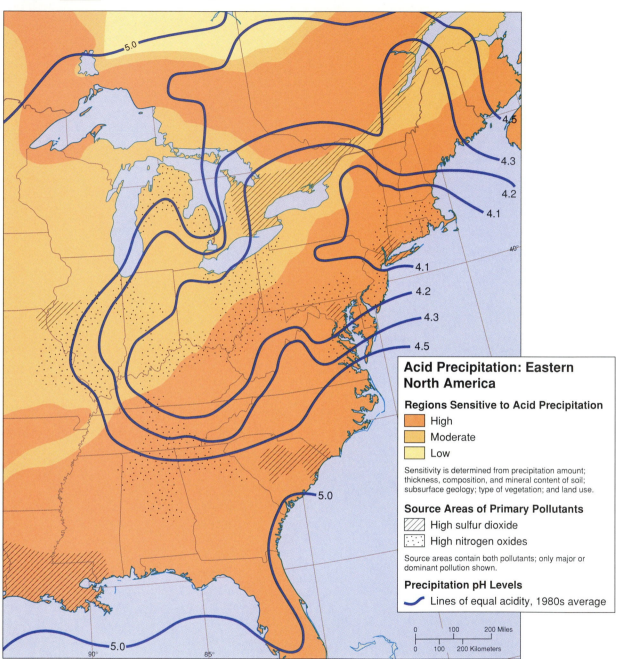

Acid Precipitation: Eastern North America

Regions Sensitive to Acid Precipitation

- High
- Moderate
- Low

Sensitivity is determined from precipitation amount; thickness, composition, and mineral content of soil; subsurface geology; type of vegetation; and land use.

Source Areas of Primary Pollutants

- High sulfur dioxide
- High nitrogen oxides

Source areas contain both pollutants; only major or dominant pollution shown.

Precipitation pH Levels

- Lines of equal acidity, 1980s average

| 0 | 100 | 200 Miles |
| 0 | 100 | 200 Kilometers |

Acid deposition is the consequence of the combination of atmospheric water vapor with sulfur oxides (mostly from coal-burning electric power generators and from the iron and steel industry) or with nitrogen oxides (mainly from utilities and automobiles). As the gases travel with the wind and combine with atmospheric water, they form sulfuric and nitric acids that pollute both local and far-distant regions. Atmospheric acid levels are shown on the map in terms of the pH scale; the lower the number on this scale, the greater the acid present. In the eastern parts of North America, the highest atmospheric acidity comes mainly from coal-fired electric utilities and iron and steel manufacturing plants upwind from the greatest concentrations of atmospheric acidity. Acids in the atmosphere will fall with precipitation as acid rain or snow, often with the same pH level as lemon juice, or as dry acid fallout. Either way, acid deposition damages trees and other vegetation, increases acidity levels in the soil, and contributes to the corrosion of surfaces of buildings and statues. At least for now, acid deposition is primarily a problem of the more highly developed regions of the world.

Map 18 Major Polluters and Common Pollutants

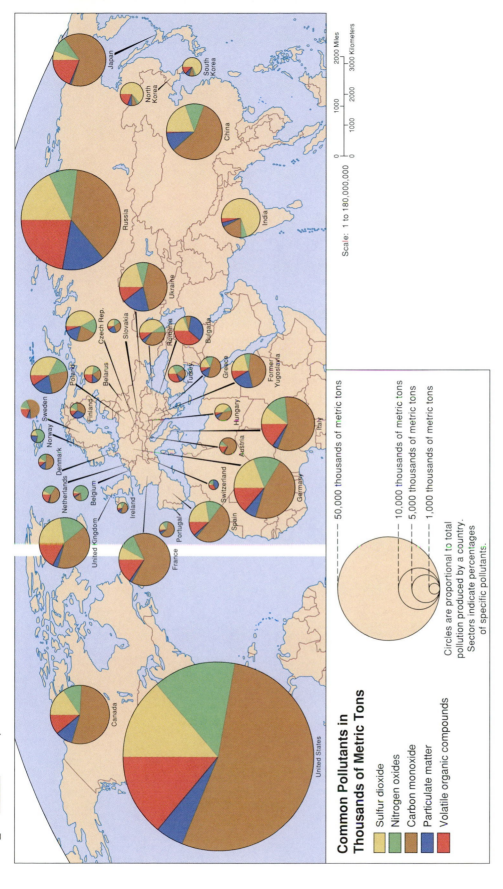

Common Pollutants in Thousands of Metric Tons

- Sulfur dioxide
- Nitrogen oxides
- Carbon monoxide
- Particulate matter
- Volatile organic compounds

50,000 thousands of metric tons

10,000 thousands of metric tons
5,000 thousands of metric tons
1,000 thousands of metric tons

Circles are proportional to total pollution produced by a country. Sectors indicate percentages of specific pollutants.

Scale: 1 to 180,000,000

0 1000 2000 Miles

0 1000 2000 3000 Kilometers

More than 90 percent of the world's total of anthropogenic (human-generated) air pollutants come from the heavily populated industrial regions of North America, Europe, South Asia (primarily in India) and East Asia (mainly in China, Japan, and the two Koreas). This map shows the origins of the five most common pollutants: sulfur dioxide, nitrogen oxide, carbon monoxide, particulate matter, and volatile organic compounds. These substances are produced both by industry and by the combustion of fossil fuels that generate electricity and power trains, planes, automobiles, buses, and trucks. In addition to combining with other components of the atmosphere and with one another to produce smog, they are the chief ingredients in acid accumulations in the atmosphere, which ultimately result in acid deposition, either as acid precipitation or dry acid fallout. Like other forms of pollutants, these air pollutants do not recognize political boundaries, and regions downwind of major polluters receive large quantities of pollutants from areas over which they often have no control.

Map 19 The Ozone Hole: Antarctica, October 1989–October 1994

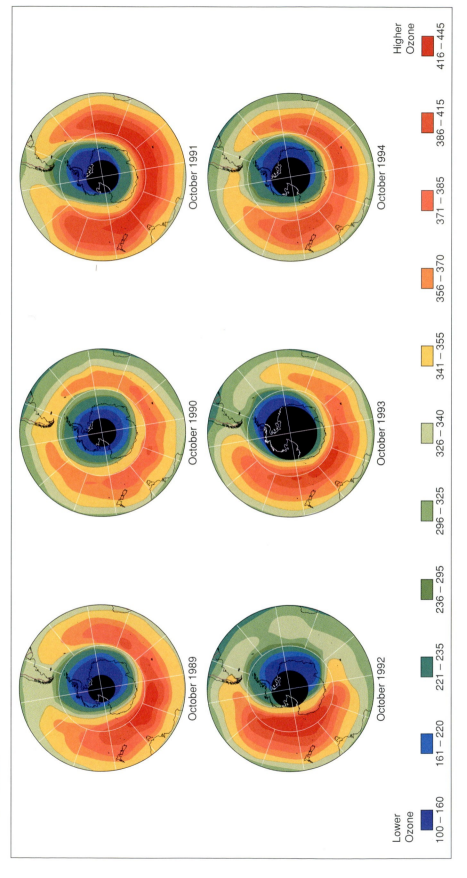

Lower Ozone

| 100 – 160 | 161 – 220 | 221 – 235 | 236 – 295 | 296 – 325 | 326 – 340 | 341 – 355 | 356 – 370 | 371 – 385 | 386 – 415 | 416 – 445 |

Higher Ozone

October 1989

October 1990

October 1991

October 1992

October 1993

October 1994

In that portion of the atmosphere called the stratosphere, at elevations above 35,000 feet or 10,000 meters, is a layer in the atmosphere naturally rich in ozone, a form of oxygen, produced by sunlight interacting with normal atmospheric oxygen. This "ozone layer" performs a number of important functions for the global ecosystem, not the least of which is screening out large amounts of potentially harmful ultraviolet radiation. However, the ozone layer has been reduced by a chemical interaction between ozone and human-produced compounds, particularly chlorofluorocarbons or CFCs. CFCs were widely employed as coolants for refrigerators and air condition-ers and as propellants in spray cans prior to international agreements banning their continued manufacture. CFCs continue to be used (often illegally) in some countries. The use of CFCs has caused a hole to develop in the ozone layer, centered over the South Pole. Scientists fear that this "ozone hole" will continue to grow larger until the ozone layer will offer little (or no) protection against ultraviolet radiation. The short-term consequences for humans are genetic damage and increased skin cancers. The long-term consequences could be more dramatic, involving widespread disruption of current vegetation patterns and global climate change.

Map 20 Global Carbon Dioxide Emissions

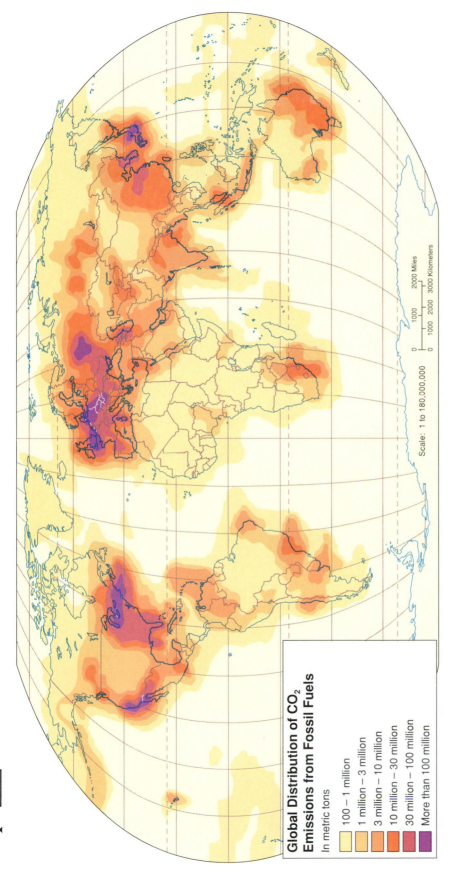

Global Distribution of CO₂ Emissions from Fossil Fuels

In metric tons

- 100 – 1 million
- 1 million – 3 million
- 3 million – 10 million
- 10 million – 30 million
- 30 million – 100 million
- More than 100 million

Scale: 1 to 180,000,000

0   1000   2000 Miles
0   1000   2000   3000 Kilometers

One of the most important components of the atmosphere is the gas carbon dioxide, the byproduct of animal respiration, of decomposition, and of combustion. During the past 200 years, atmospheric CO₂ has risen dramatically, largely as the result of the tremendous increase in fossil fuel combustion brought on by the industrialization of the world's economy and the burning and clearing of forests by the expansion of farming. While carbon dioxide by itself is relatively harmless, it is an important "greenhouse gas." The gases in the atmosphere act like the panes of glass in a greenhouse roof, allowing light in but preventing heat from escaping. The greenhouse ca-

pacity of the atmosphere is crucial for organic life and is a purely natural component of the global energy cycle. But too much carbon dioxide and other greenhouse gases such as methane could cause the earth's atmosphere to warm up too much, producing the global warming that atmospheric scientists are concerned about. Researchers estimate that if greenhouse gases such as carbon dioxide continue to increase at their present rates, the earth's mean temperature could rise between 1.5 and 4.5 degrees Celsius by the middle of the next century. Such a rise in global temperatures would produce massive alterations in the world's climate patterns.

Map 21 Potential Global Temperature Change

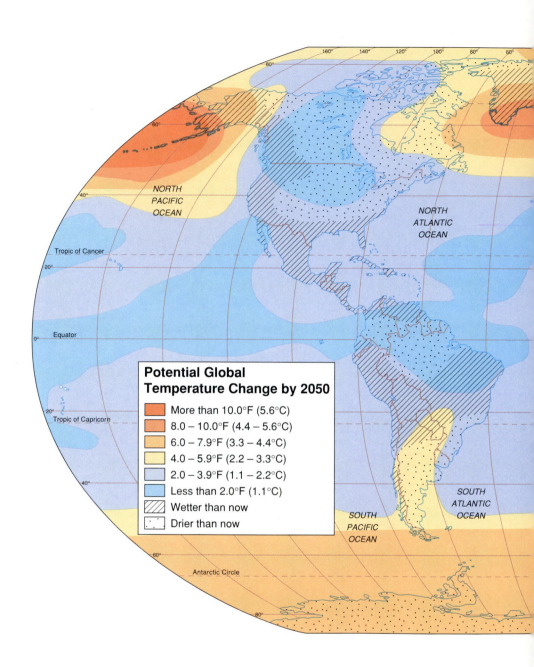

Potential Global Temperature Change by 2050

- More than 10.0°F (5.6°C)
- 8.0 – 10.0°F (4.4 – 5.6°C)
- 6.0 – 7.9°F (3.3 – 4.4°C)
- 4.0 – 5.9°F (2.2 – 3.3°C)
- 2.0 – 3.9°F (1.1 – 2.2°C)
- Less than 2.0°F (1.1°C)
- Wetter than now
- Drier than now

NORTH PACIFIC OCEAN

NORTH ATLANTIC OCEAN

Tropic of Cancer

Equator

Tropic of Capricorn

SOUTH PACIFIC OCEAN

SOUTH ATLANTIC OCEAN

Antarctic Circle

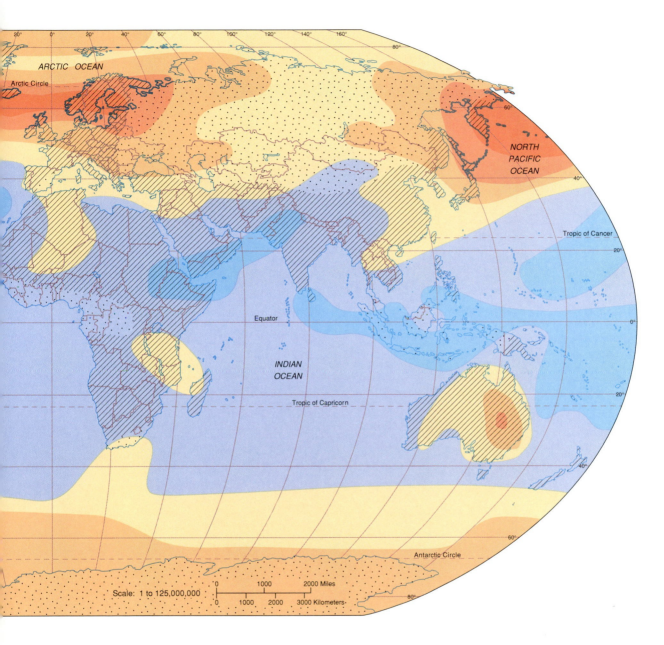

According to most atmospheric scientists, one of the major problems of the twenty-first century is likely to be increasing global temperatures. These increasing temperatures, it is believed, will be produced as the atmosphere's natural ability to trap and retain heat is enhanced by increased percentages of carbon dioxide and other "greenhouse gases" in the atmosphere. Not all scientists believe that global warming will take place. But computer models based on atmospheric percentages of carbon dioxide resulting from present use of fossil fuels show that warming is a possibility. Increased temperatures would cause precipitation patterns to change fairly dramatically and would produce a number of other harmful effects, including a rise in the level of the world's oceans that could flood most coastal cities.

Map 22 Reliance on Nuclear Power

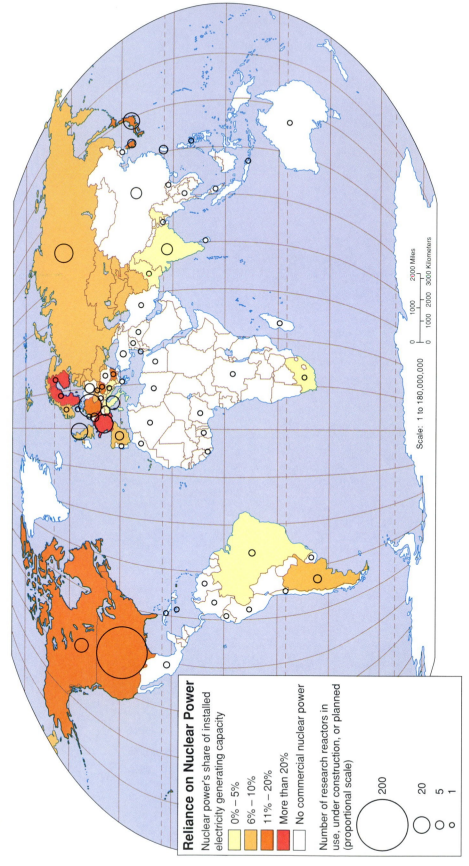

Reliance on Nuclear Power

Nuclear power's share of installed
electricity generating capacity

- 0% – 5%
- 6% – 10%
- 11% – 20%
- More than 20%
- No commercial nuclear power

Number of research reactors in
use, under construction, or planned
(proportional scale)

200
20
5
1

Scale: 1 to 180,000,000

2000 Miles
3000 Kilometers
1000   2000
0   1000   2000
0

Controlled nuclear fission is an energy source widely used in the developed nations of the world. Using refined uranium as a fuel source, nuclear reactors produce enormous quantities of heat energy, which is then used to produce steam to drive turbines, in turn producing electricity. Since nuclear power generation relies upon some of the same technologies used in producing fission weapons for warfare, people are concerned about the safety of nuclear power and about the disposal of spent but still dangerous radioactive waste materials. While the nuclear power industry has a generally good safety record in Europe and North America, severe problems have occurred in Russia and its neighbors, where reactor technology differs from that in the West.

Map 23 The Chernobyl Nuclear Accident

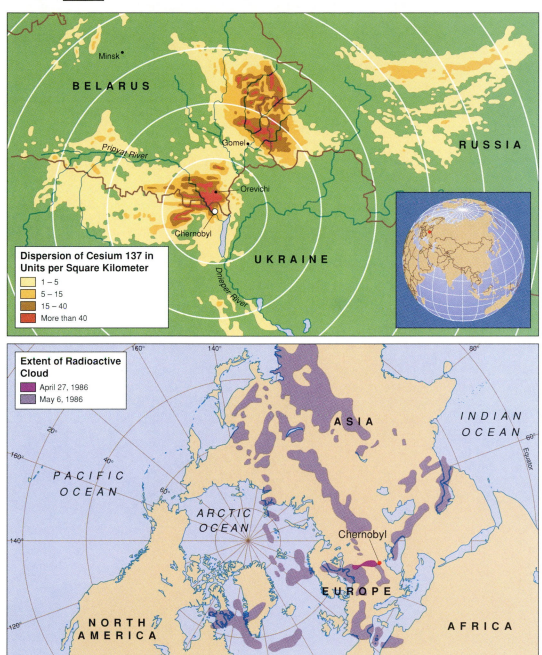

Dispersion of Cesium 137 in Units per Square Kilometer
- 1 – 5
- 5 – 15
- 15 – 40
- More than 40

BELARUS

Minsk

RUSSIA

Pripyat River

Gomel

Orevichi

Chernobyl

UKRAINE

Dnieper River

Extent of Radioactive Cloud
- April 27, 1986
- May 6, 1986

ASIA

INDIAN OCEAN

PACIFIC OCEAN

Equator

ARCTIC OCEAN

Chernobyl

EUROPE

NORTH AMERICA

AFRICA

Note: Place names are as of 1996.

In April 1986 a major accident occurred at one of the four nuclear reactors near Chernobyl, a city in what was then the Ukrainian Republic of the U.S.S.R., near where the present boundaries of the countries of Belarus, Ukraine, and Russia meet. An equipment failure during a test of the facility's shutdown procedures caused a hydrogen explosion, which resulted in the release of high levels of the radioactive element cesium 137 into the air. This radioactive element polluted both soil and water in the plant's vicinity and forced a general evacuation of the area. Cesium 137 was also spread widely through the western parts of the Soviet Union, as meat and dairy products from the Chernobyl region were marketed and consumed well after the accident. The cost in human life in the immediate vicinity of the accident has been fairly well documented. What is more difficult to assess are the long-term consequences of the consumption of cesium 137–tainted products and the worldwide dispersion of a radioactive cloud that eventually brought increased levels of radioactivity to every continent except Australia and Antarctica.

Table E

Greenhouse Gas Emissions, 1992[1] (in thousands of metric tons)

	INDUSTRIAL CO_2[2]	LAND USE CO_2[3]	METHANE[4]	PER CAPITA CO_2
WORLD	22,339,408	4,100,000	270,000	4.10
AFRICA	**715,773**	**730,000**	**21,000**	X
Algeria	79,712	6,900	1,700	3.00
Angola	4,525	16,000	340	0.44
Benin	612	3,200	65	0.11
Botswana	2,173	3,200	110	1.65
Burkina Faso	557	3,400	260	0.07
Burundi	191	130	33	0.04
Cameroon	2,231	28,000	260	0.18
Central African Republic	216	23,000	110	0.07
Chad	253	7,100	240	0.04
Congo	3,972	14,000	31	1.69
Egypt	83,997	X	1,000	1.54
Equatorial Guinea	117	3,000	2	0.33
Eritrea	X	X	68	X
Ethiopia	2,906	8,000	1,200	0.04
Gabon	5,569	51,000	250	4.51
Gambia	198	130	20	0.22
Ghana	3,781	18,000	130	0.22
Guinea	1,026	10,000	520	0.18
Guinea-Bissau	209	1,800	59	0.22
Ivory Coast	6,309	15,000	110	0.48
Kenya	5,342	1,400	540	0.22
Lesotho	X	X	44	X
Liberia	278	9,600	31	0.11
Libya	39,520	76	480	8.10
Madagascar	945	21,000	860	0.07
Malawi	652	11,000	72	0.07
Mali	443	8,400	350	0.04
Mauritania	2,869	1	140	1.36
Mauritius	1,356	9	5	1.25
Morocco	27,344	4,700	360	1.03
Mozambique	997	15,000	98	0.07
Namibia	X	1,800	96	X
Niger	1,085	X	160	0.15
Nigeria	96,513	24,000	4,500	0.84
Rwanda	451	170	34	0.07
Senegal	2,810	4,700	190	0.37
Sierra Leone	432	1,800	150	0.11
Somalia	15	430	540	0.00
South Africa	290,291	14,000	2,400	7.29
Sudan	3,462	38,000	1,100	0.15
Swaziland	267	370	25	0.33
Tanzania	2,103	22,000	760	0.07
Togo	733	2,100	45	0.18
Tunisia	13,560	640	130	1.61
Uganda	953	5,000	250	0.04
Zaire	4,181	280,000	380	0.11
Zambia	2,481	34,000	150	0.29
Zimbabwe	18,675	5,300	230	1.76
NORTH AND MIDDLE AMERICA	**5,715,466**	**190,000**	**35,000**	X
Belize	264	980	4	1.32
Canada	409,862	X	3,500	14.99
Costa Rica	3,807	14,000	130	1.21
Cuba	28,623	3,200	370	2.64
Dominican Republic	10,248	4,800	210	1.36
El Salvador	3,550	520	88	0.66
Guatemala	5,657	21,000	160	0.59
Haiti	784	470	88	0.11
Honduras	3,059	19,000	130	0.55
Jamaica	8,042	7,200	31	3.26
Mexico	332,852	63,000	3,100	3.77

(Continued on next page)

	INDUSTRIAL CO_2[2]	LAND USE CO_2[3]	METHANE[4]	PER CAPITA CO_2
Nicaragua	2,495	33,000	140	0.82
Panama	4,228	21,000	100	1.69
Trinidad and Tobango	20,643	1,200	300	16.30
United States	4,881,349	X	27,000	19.13
SOUTH AMERICA	**605,029**	**1,800,000**	**140,000**	**X**
Argentina	117,003	85,000	3,700	3.52
Bolivia	6,632	140,000	540	0.88
Brazil	217,074	1,100,000	9,900	1.39
Chile	34,738	33,000	380	2.56
Colombia	61,493	110,000	2,100	1.83
Ecuador	18,888	72,000	490	1.72
Guyana	835	6,900	31	1.03
Paraguay	2,620	35,000	440	0.59
Peru	22,277	96,000	520	0.99
Suriname	2,008	5,100	42	4.58
Uruguay	5,038	1,300	690	1.61
Venezuela	116,424	170,000	2,000	5.75
ASIA	**7,118,317**	**1,300,000**	**140,000**	**X**
Afghanistan	1,392	1,100	210	0.07
Armenia	4,199	X	42	1.21
Azerbaijan	63,878	X	450	8.76
Bangladesh	17,217	7,700	3,900	0.15
Bhutan	132	4,500	36	0.07
Cambodia	476	35,000	140	0.04
China	2,667,982	150,000	47,000	2.27
Georgia	13,839	X	130	2.53
India	769,440	65,000	33,000	0.88
Indonesia	184,585	410,000	10,000	0.95
Iran	235,478	10,000	3,300	3.81
Iraq	64,527	24	540	3.33
Israel	41,605	X	73	8.10
Japan	1,093,470	X	3,900	8.79
Jordan	11,311	97	50	2.64
Kazakhstan	297,982	X	2,500	17.48
Korea, North	252,750	700	1,600	11.21
Korea, South	289,833	1,500	1,400	6.56
Kuwait	15,971	X	170	8.10
Kyrgyzstan	X	X	190	X
Laos	271	X	270	0.07
Lebanon	11,051	91	29	3.88
Malaysia	70,492	210,000	960	3.74
Mongolia	9,281	480	300	4.03
Myanmar	4,386	130,000	2,300	0.11
Nepal	1,297	9,000	610	0.07
Oman	10,036	X	150	6.12
Pakistan	71,902	14,000	3,300	0.59
Phillippines	49,698	110,000	1,900	0.77
Saudi Arabia	220,620	60	2,600	13.85
Singapore	49,790	X	65	17.99
Sri Lanka	4,972	4,300	610	0.29
Syria	42,407	790	620	3.19
Tajikistan	3,972	X	150	0.70
Thailand	112,477	92,000	5,500	2.02
Turkey	145,490	X	1,100	2.49
Turkmenistan	42,257	X	1,000	10.96
United Arab Emirates	70,616	X	520	42.28
Uzbekistan	123,253	X	1,300	5.75
Vietnam	21,522	40,000	4,400	0.29
Yemen	10,083	X	90	0.81
EUROPE	**6,866,494**	**11,000**	**53,000**	**X**
Albania	3,968	X	92	1.21
Austria	56,572	X	260	7.29
Belarus	102,028	X	510	9.89
Belgium	101,768	X	190	10.19
Bosnia-Herzegovina	15,055	X	38	3.37

(Continued on next page)

	INDUSTRIAL CO_2[2]	LAND USE CO_2[3]	METHANE[4]	PER CAPITA CO_2
Bulgaria	54,359	X	8,800	6.08
Croatia	16,210	X	73	3.33
Czech Republic	135,608	X	380	13.04
Denmark	53,897	X	260	10.44
Estonia	20,885	X	59	13.19
Finland	41,176	X	150	8.21
France	362,076	X	1,800	6.34
Germany	878,136	X	3,400	10.96
Greece	73,859	X	330	7.25
Hungary	59,910	X	270	5.72
Iceland	1,777	X	15	6.85
Ireland	30,851	X	520	8.87
Italy	407,701	X	1,500	7.03
Latvia	14,781	X	100	5.53
Lithuania	22,006	X	150	5.86
Macedonia	4,100	X	40	1.98
Moldova	14,209	X	110	3.26
Netherlands	139,027	X	1,400	9.16
Norway	60,247	X	2,400	14.03
Poland	341,892	X	1,800	8.90
Portugal	47,181	X	170	4.80
Romania	122,103	X	840	5.24
Russia	2,103,132	X	17,000	14.11
Slovakia	36,999	X	320	7
Slovenia	5,503	X	35	2.75
Spain	223,196	X	1,400	5.72
Sweden	56,796	X	200	6.56
Switzerland	43,701	X	140	6.38
Ukraine	611,342	X	3,600	11.72
United Kingdom	566,246	X	3,800	9.78
Yugoslavia	38,197	X	180	3.63
OCEANIA	**297,246**	**38,000**	**5,800**	**X**
Australia	267,937	X	4,800	15.24
Fiji	711	1,400	25	0.95
New Zealand	26,179	X	1,000	7.58
Papua New Guinea	2,257	35,000	13	0.55
Solomon Islands	161	1,800	1	0.48

[1]Greenhouse gases are those gases, occurring either naturally or through human activities, that enhance the ability of the earth's atmosphere to trap and retain heat energy. Heat energy is solar energy that has been absorbed by the earth's land and water surfaces, converted from light energy to long-wave or heat energy, and radiated back to warm the atmosphere. Many atmospheric scientists believe that an increase in the atmospheric content of greenhouse gases through fossil-fuel burning, forest clearance, and other anthropogenic processes may cause global warming.

[2]Industrial carbon dioxide results from the combustion of solid, liquid, and gas fuels, gas flaring during petroleum extraction, and cement manufacture.

[3]Land use carbon dioxide generation is produced by land use changes that create higher than normal emissions; chief among these is forest clearance by burning, but the category would also include wetlands restoration, irrigation agriculture, and livestock feeding.

[4]Methane (CH_4) is produced chiefly from oil and natural gas extraction and distribution, coal mining, wetland rice agriculture, municipal solid waste decomposition, and livestock.

Source: World Resources 1996–97.

Part IV

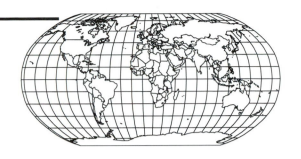

Human Impact on Fresh Water and the Oceans

Map 24 World Water Resources: Availability of Renewable Water Per Capita

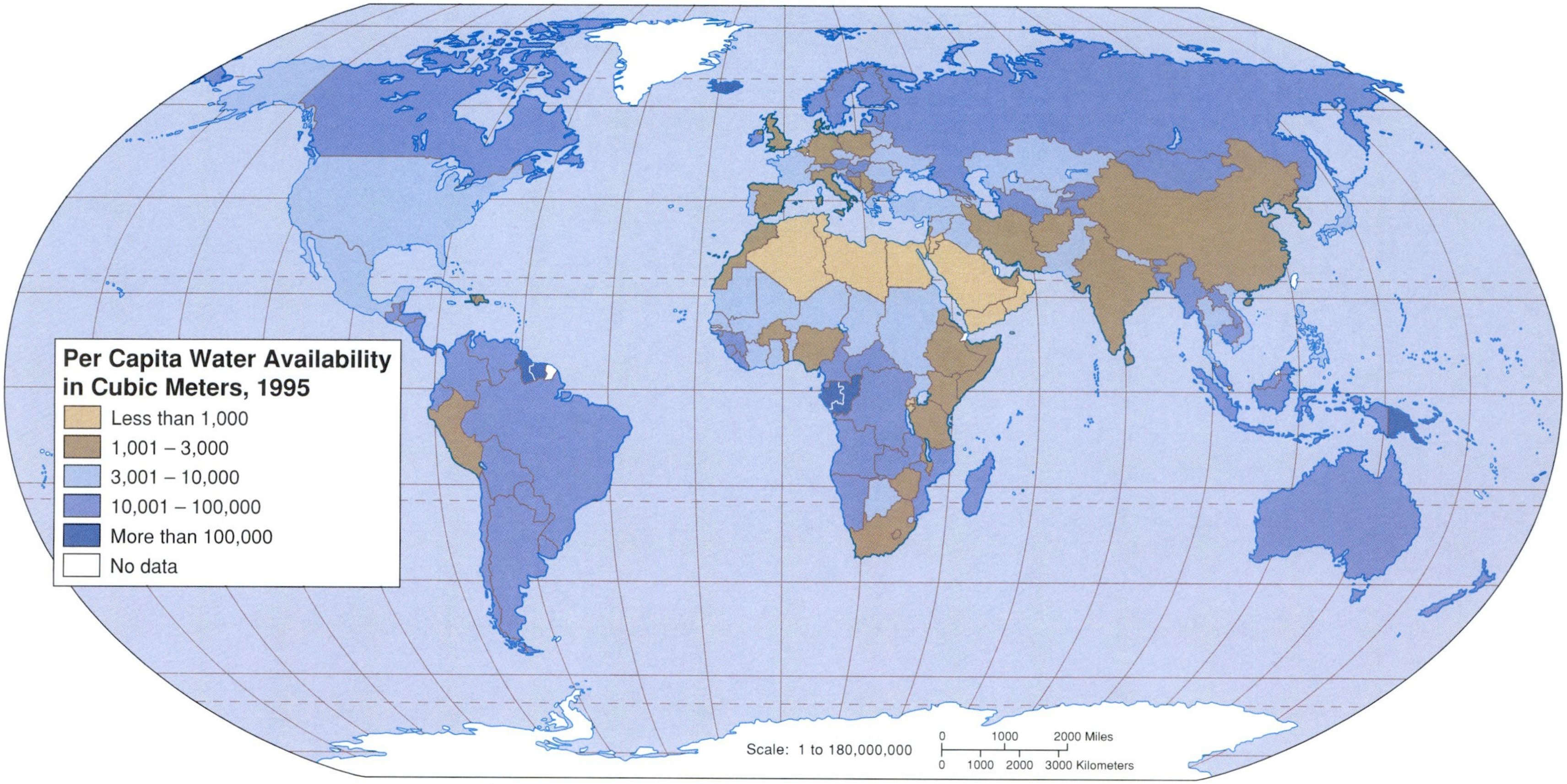

Per Capita Water Availability in Cubic Meters, 1995

- Less than 1,000
- 1,001 – 3,000
- 3,001 – 10,000
- 10,001 – 100,000
- More than 100,000
- No data

Scale: 1 to 180,000,000

```
0        1000        2000 Miles
0    1000    2000    3000 Kilometers
```

Renewable water resources are usually defined as the total water available from streams and rivers (including flows from other countries), ponds and lakes, and groundwater storage or aquifers. Not included in the total of renewable water would be water that comes from such nonrenewable sources as desalinization plants or melted icebergs. While the concept of renewable or flow resources is a traditional one in resource management, in fact, few resources, including water, are truly renewable when their use is excessive. The water resources shown here are indications of that principle. A country like the United States possesses truly enormous quantities of water. But the United States also uses enormous quantities of water. The result is availability of renewable water that is—largely because of excess use—much less than in many other parts of the world where the total supply of water is significantly less.

– 58 –

Map 25 World Water Resources: Annual Withdrawal Per Capita

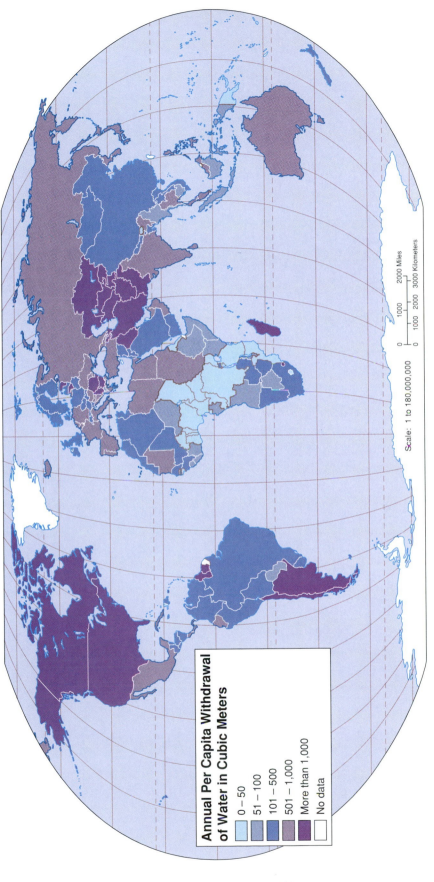

**Annual Per Capita Withdrawal
of Water in Cubic Meters**

- 0 – 50
- 51 – 100
- 101 – 500
- 501 – 1,000
- More than 1,000
- No data

Scale: 1 to 180,000,000

0   1000   2000 Miles

0   1000   2000   3000 Kilometers

Water resources must be viewed like a bank account in which deposits and withdrawals are made. As long as the deposits are greater than the withdrawals, a positive balance remains. But when the withdrawals begin to exceed the deposits, sooner or later (depending on the relative sizes of the deposits and withdrawals) the account becomes overdrawn. For many of the world's countries, annual availability of water is insufficient to cover the demand. In these countries, reserves stored in groundwater are being tapped, resulting in depletion of the water supply (think of this as shifting money from a savings account to a checking account). The water supply can maintain its status as a renewable resource only if deposits continue to be greater than withdrawals, and that seldom happens. In general, countries with high levels of economic development and countries that rely on irrigation agriculture are the most spendthrift when it comes to their water supplies.

– 59 –

Map 26 World Water Supplies

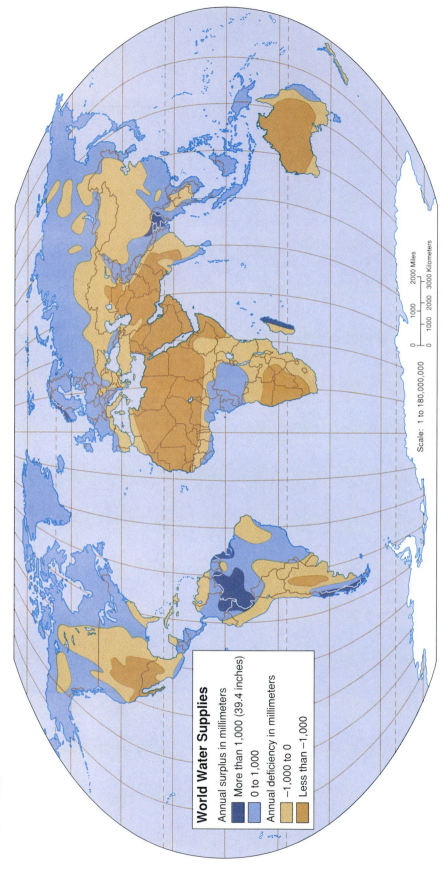

Scale: 1 to 180,000,000

0   1000   2000 Miles
0 1000 2000 3000 Kilometers

World Water Supplies

Annual surplus in millimeters
- More than 1,000 (39.4 inches)
- 0 to 1,000

Annual deficiency in millimeters
- −1,000 to 0
- Less than −1,000

The supply of groundwater, surface water, and soil moisture is largely dependent upon the natural surplus or deficit of water. Water surpluses exist in those areas where water income, in the form of precipitation, exceeds the loss of water through evaporation from the soil, from ponds, lakes, and streams, and through transpiration, the evaporation of water from leaf surfaces of plants. Both evaporation and transpiration (combined as evapotranspiration) are functions of temperature, with higher temperatures producing higher rates of evapotranspiration. Regions of water deficit are particularly susceptible to soil chemistry changes because farming in water-poor areas requires irrigation. Regions of water surplus may be vulnerable to water pollution as abundant runoff combines with polluting substances and washes them into ponds and lakes.

Map 27 Groundwater Use: The Oglalla Aquifer

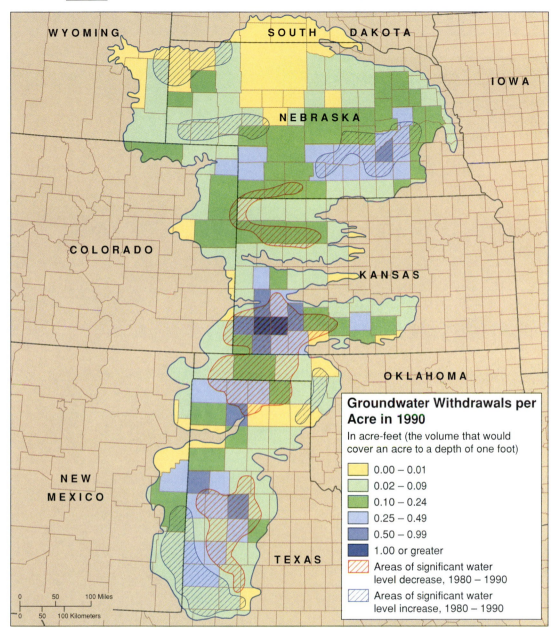

Groundwater Withdrawals per Acre in 1990

In acre-feet (the volume that would cover an acre to a depth of one foot)

- 0.00 – 0.01
- 0.02 – 0.09
- 0.10 – 0.24
- 0.25 – 0.49
- 0.50 – 0.99
- 1.00 or greater
- Areas of significant water level decrease, 1980 – 1990
- Areas of significant water level increase, 1980 – 1990

An aquifer is an underground water source, usually consisting of a porous medium of gravel, sand, and bedrock, filled with water trapped by a layer of impermeable stone. The aquifer may feed rivers and lakes through springs as water seeps to the surface, and water is also withdrawn from the aquifer for human uses, particularly irrigation. The aquifer is recharged by precipitation and stream runoff. Most aquifers represent thousands of years of accumulation of stored water. The Oglalla aquifer is the most important aquifer for American agriculture, as it underlies the Great Plains and has allowed the expansion of intensive American agriculture into that primary agricultural region. When aquifers such as the Oglalla are tapped for irrigation, however, the rates of withdrawal far exceed the rates of recharge. Since the beginning of major use of the Oglalla aquifer in the 1940s, the average water level has dropped more than 10 feet, with drops of more than 100 feet in the Texas panhandle. Although Nebraska has by far the greatest amount of water, even there the aquifer is dropping steadily as the result of withdrawal rates that amount to 30 percent of the total U.S. groundwater used for irrigation. Although resource managers have often thought of groundwater as a renewable or "flow" resource, in fact it is just as nonrenewable in the human time frame as an oil field or iron mine. When the Oglalla aquifer is gone, it will take tens of thousands of years to restore it.

Map 28 Pollution of the Oceans

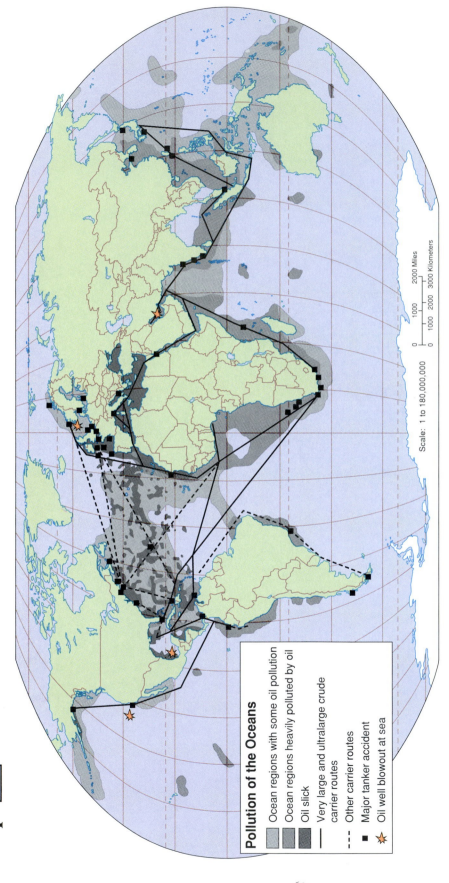

Pollution of the Oceans

- Ocean regions with some oil pollution
- Ocean regions heavily polluted by oil
- Oil slick
- Very large and ultralarge crude carrier routes
- Other carrier routes
- Major tanker accident
- Oil well blowout at sea

Scale: 1 to 180,000,000

0	1000	2000 Miles
0	1000 2000	3000 Kilometers

The pollution of the world's oceans has long been a matter of concern to environmental scientists. The great circulation systems of the ocean (shown in Map 4) are one of the controlling factors of the earth's natural environment, and modifications to those systems have unknown consequences. This map is based on what we can measure: (1) areas of oceans where oil pollution has been proven to have inflicted significant damage to ocean ecosystems and life-forms (including phytoplankton, the oceans' primary food producers, equivalent to land-based vegetation) and (2) areas of oceans where unusually high concentrations of hydrocarbons from oil spills may have inflicted some damage to the oceans' biota. A glance at the map shows that there are few areas of the world's oceans where some form of pollution is not a part of the environmental system. What the map does not show in detail, because of the scale, are the dramatic consequences of large individual pollution events: the wreck of the *Exxon Valdez* and the polluting of Prince William Sound or the environmental devastation produced by the Gulf War in the Persian Gulf.

Map 29 Oil Spill Disaster I: Prince William Sound

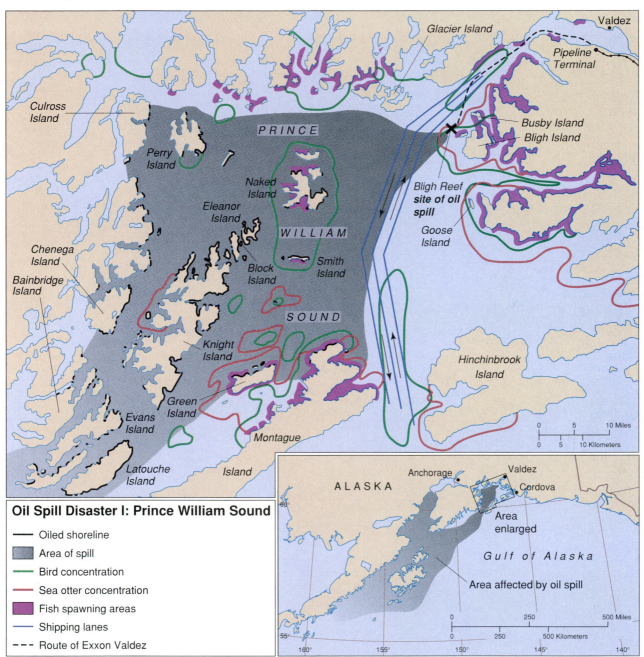

Oil Spill Disaster I: Prince William Sound

— Oiled shoreline

▉ Area of spill

— Bird concentration

— Sea otter concentration

▉ Fish spawning areas

— Shipping lanes

--- Route of Exxon Valdez

On Friday, March 24, 1989, the supertanker *Exxon Valdez,* filled with oil loaded from the terminus of the Alaska Pipeline near Valdez, Alaska, ran aground on Bligh Reef at the head of Prince William Sound. As the ship breached, it poured millions of gallons of oil into the sound, one of the world's most productive and pristine high-latitude coastal ecosystems and a prime breeding ground for sea otter, herring, salmon, and countless sea birds. The oil, floating on the sound's surface, was transported by winds and currents throughout the sound, fouling some beaches and shorelines while leaving others untouched. More than 1,200 miles of shoreline were even-

tually polluted by oil, and the loss to wildlife was incalculable. The spill affected four national wildlife refuges, one national forest, and three national parks. To give some perspective to the magnitude of the oil slick that spread from the wrecked ship, if it had occurred on the Atlantic seaboard of the United States, the slick would have stretched from Cape Cod to the Outer Banks of North Carolina. Although the most advanced sensing devices were used to trace the spread of oil and modern recovery methods were used to contain it, it will take decades for the shores of Prince William Sound to recover from this environmental disaster.

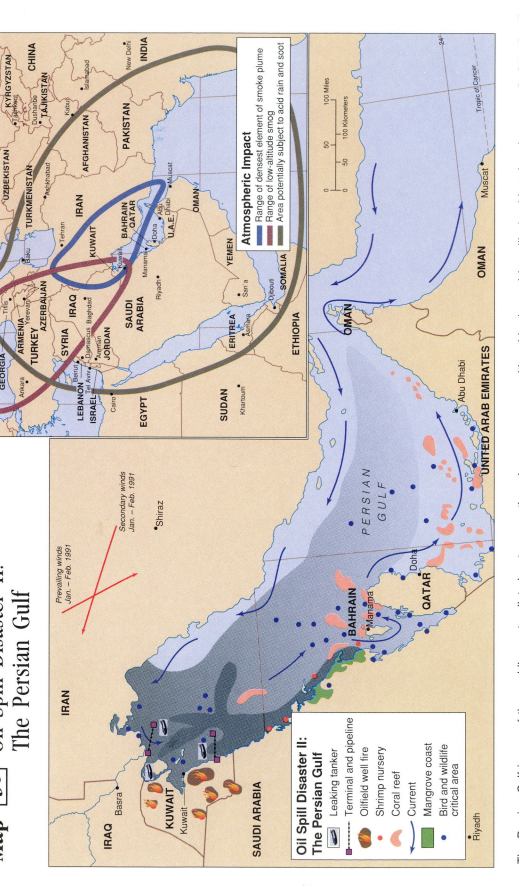

Map 30 Oil Spill Disaster II: The Persian Gulf

Atmospheric Impact
- Range of densest element of smoke plume
- Range of low-altitude smog
- Area potentially subject to acid rain and soot

Oil Spill Disaster II: The Persian Gulf
- Leaking tanker
- Terminal and pipeline
- Oilfield well fire
- Shrimp nursery
- Coral reef
- Current
- Mangrove coast
- Bird and wildlife critical area

Prevailing winds Jan. – Feb. 1991

Secondary winds Jan. – Feb. 1991

The Persian Gulf is one of the world's most polluted waterways, the price of being the oil highway for Middle East oil bound for the ports of Europe, Japan, and North America. Each year, "normal" seepage from tankers and terminals adds 250,000 barrels of oil to the gulf's waters. In spite of this pollution, the western gulf is a haven for wildlife and aquatic life, supporting rich mangrove coasts and abundant habitats for birds, turtles, and the dugong (a marine mammal). During the 1991 Gulf War, however, military operations resulted in more than 500 oil well fires that released enormous clouds of smoke transported by winds as far east as India, threatening crops with acid rain, and in spills (accidental and, apparently, intentional) from offshore oil terminals and from leaking tankers damaged during military action. The counterclockwise circulation of water in the gulf kept oil damage from the Iranian shore but concentrated it along the western shore, fouling nearly all the shoreline of Kuwait and Saudi Arabia. It also damaged the most sensitive biological areas of the entire gulf region. At its farthest extent, the oil slick spread from the Sea Island Terminal off the Kuwait shore to the north shore of Bahrain; it represented one of the most massive contamination events of the century.

Map 31 Estuarine Pollution: Chesapeake Bay

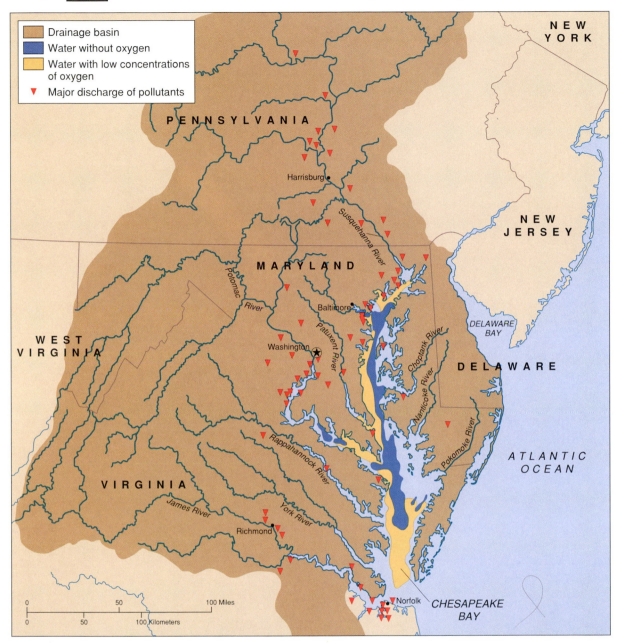

Drainage basin
Water without oxygen
Water with low concentrations of oxygen
▼ Major discharge of pollutants

Of all the ecosystems outside the tropics, the most productive are the estuarine systems, where freshwater mingles with saltwater and where terrestrial ecosystems merge with aquatic ones. The most productive of these abundant natural producers of food is North America's Chesapeake Bay, the submerged lower end of the great glacial river valley of the Susquehanna River, now fed not only by the Susquehanna but by the Patuxent, Potomac, York, and James Rivers and a host of smaller streams as well. Unfortunately, the Chesapeake region's abundance has meant overuse of its resources. Particularly stressed has been the natural purification function of running water. Because runoff from the land proceeds rapidly to the sea through the Chesapeake, industrial and agricultural sources of pollution have crowded the bay's coastal zone, utilizing the outflow to carry away from the land the effluvia of industrial agriculture and manufacturing systems. The role of the bay as a transportation nexus has added to the water quality problem through pollution from oil spills and seepage from tankers carrying Middle Eastern, Venezuelan, and Mexican oil to cities at the bay's head. While a cleanup effort has been under way for over 12 years and some of the bay's rich fisheries have begun to recover, the waters of the Chesapeake were so fouled by decades of abuse that they will take decades to recover.

Map 32 Lake Pollution: The Great Lakes

Great Lakes drainage basin
● Most polluted areas, according to the Great Lakes Water Quality Board
● Other "hot spots" of toxic concentrations in water and sediments
▲ U.S. Superfund sites (not including those added after October 1983)
▲ Canadian industrial waste sites identified by the Ontario Ministry of the Environment as "needing monitoring"
Eutrophic areas

The Great Lakes of the United States and Canada are one of the world's most unique freshwater systems. Occupying river valleys scoured deeply by glacial action during the Pleistocene Era and then filled with meltwater as the glaciers receded, the Great Lakes contain over 25 percent of the world's supply of freshwater. They have been used as a source of food, a medium of transport, and a receiver of industrial wastes for over a century. The lower lakes—in particular Michigan, Erie, and Ontario—occupy a region that is both the heartland of North American agriculture and a major portion of the North American core industrial area. Stretching from Chicago to Montreal, this great agricultural and industrial region has made enormous demands upon the lakes: to carry iron ore and coal to the steel mills of Gary, Indiana; to carry the finished products of Middle American industry to the wider world at large; and to carry off the unwanted by-products of the agricultural/industrial core as well. Rivers throughout the Great Lakes drainage basin, subjected to pollution from industrial and agricultural sources, were so fouled that the lower lakes, Erie in particular, had achieved the status of "dead lakes" by the end of the 1960s. Indeed, the Cuyahoga River, flowing through Cleveland into Lake Erie, epitomized the pollution problem when it actually caught on fire. Floating hydrocarbon wastes from the upstream synthetic rubber factories ignited, and the flames reached the height of a five-story building before being brought under control. Since the Cuyahoga fire, efforts to clean up the polluted lakes have been generally successful. Lake Erie once again has live fish, and the future of the lakes looks much brighter now than it did 30 years ago.

Map 33 A Declining Food Supply? The World's Marine and Freshwater Systems

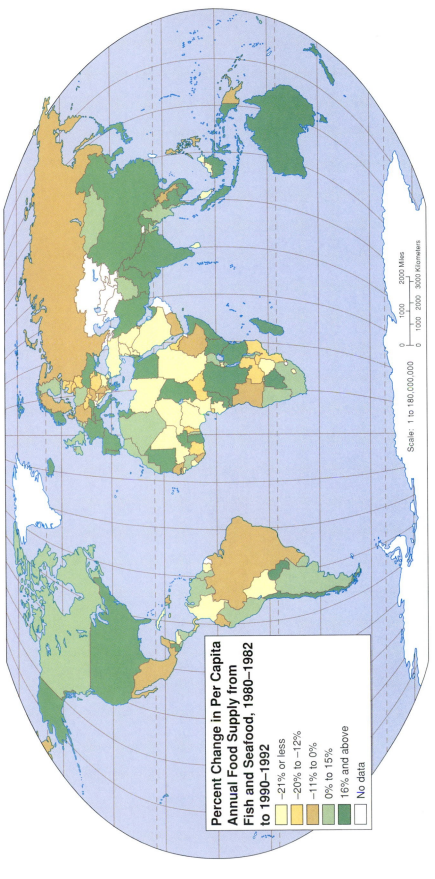

Percent Change in Per Capita Annual Food Supply from Fish and Seafood, 1980–1982 to 1990–1992

- −21% or less
- −20% to −12%
- −11% to 0%
- 0% to 15%
- 16% and above
- No data

Scale: 1 to 180,000,000

0   1000   2000 Miles

0   1000 2000 3000 Kilometers

Not that many years ago, food supply experts were confidently predicting that the "starving millions" of the world of the future could be fed from the unending bounty of the world's oceans. While the annual catch from the sea helped to keep hunger at bay for a time, by the late 1980s it had become apparent that without serious human intervention in the form of aquaculture, the supply of fish would not be sufficient to offset the population/food imbalance that was beginning to affect so many of the world's regions. The development of factory-fishing with advanced equipment to locate fish and process them before they went to market increased the supply of food from the ocean, but in that increase was sown the seeds of future problems. The factory-fishing system, efficient in terms of economics, was costly in terms of fish populations. In some well-fished areas, the stock of fish that was viewed as near infinite just a few decades ago has dwindled nearly to the point of disappearance. This map shows both increases and decreases in the amount of individual countries' food supplies from the ocean. The increases are often the result of more technologically advanced fishing operations. The decreases are usually the result of the same thing.

Table F

Freshwater Resources and Withdrawals

	ANNUAL INTERNAL RENEWABLE WATER RESOURCES[a]		ANNUAL WITHDRAWALS[b]			SECTORAL WITHDRAWALS (%)[c]		
	TOTAL (cubic km)	1995 PER CAPITA (cubic meters)	TOTAL (cubic km)	PERCENTAGE OF WATER RESOURCES	PER CAPITA (cubic meters)	DOMESTIC	INDUSTRY	AGRICULTURE
WORLD	41,022	7,176	3,240.00	8	645	8	23	69
AFRICA	**3,996.0**	**5,488**	**145.14**	**4**	**199**	**7**	**5**	**88**
Algeria	14.8	528	4.50	30	180	25	15	60
Angola	184.0	16,618	0.48	0	57	14	10	76
Benin	25.8	4,770	0.15	1	28	23	10	67
Botswana	14.7	9,886	0.11	1	83	32	20	48
Burkina Faso	28.0	2,713	0.38	1	40	19	0	81
Burundi	3.6	563	0.10	3	20	36	0	64
Cameroon	268.0	20,252	0.40	0	38	46	19	35
Central African Republic	141.0	42,534	0.07	0	26	21	5	74
Chad	43.0	6,760	0.18	0	34	16	2	82
Congo	832.0	321,236	0.04	0	20	62	27	11
Egypt	58.1	923	56.40	97	956	6	9	85
Equatorial Guinea	30.0	75,000	0.01	0	15	81	13	6
Eritrea	8.8	2,492	X	X	X	X	X	X
Ethiopia	110.0	1,998	2.21	2	51	11	3	86
Gabon	164.0	124,242	0.06	0	57	72	22	6
Gambia	8.0	7,156	0.02	0	30	7	2	91
Ghana	53.2	3,048	0.30	1	35	35	13	52
Guinea	226.0	33,731	0.74	0	140	10	3	87
Guinea-Bissau	27.0	25,163	0.2	0	17	60	4	36
Ivory Coast	77.7	5,451	0.71	1	66	22	11	67
Kenya	30.2	1,069	2.05	7	87	20	4	76
Lesotho	5.2	2,551	0.05	1	30	22	22	56
Liberia	232.0	76,341	0.13	0	56	27	13	60
Libya	0.6	111	4.60	767	880	11	2	87
Madagascar	337.0	22,827	16.30	5	1,584	1	0	99
Malawi	18.7	1,678	0.94	5	86	10	3	86
Mali	67.0	6,207	1.36	2	162	2	1	97
Mauritania	11.4	5,013	163	14	923	6	2	92
Mauritius	2.2	1,979	0.36	16	410	16	7	77
Morocco	30.0	1,110	10.85	36	427	5	3	92
Mozambique	208.0	12,997	0.61	0	41	9	2	89
Namibia	45.5	29,545	0.25	1	180	29	3	68
Niger	32.5	3,552	0.50	2	69	16	2	82
Nigeria	280.0	2,506	3.63	1	41	31	15	54
Rwanda	6.3	792	0.77	12	102	5	2	94
Senegal	39.4	4,740	1.36	3	202	5	3	92
Sierra Leone	160.0	35,485	0.37	0	99	7	4	89
Somalia	13.5	1,459	0.81	6	98	3	0	97
South Africa	50.0	1,206	13.31	27	359	17	11	72
Sudan	154.0	5,481	17.80	12	633	4	1	94
Swaziland	4.5	5,275	0.66	15	1,171	2	2	96

	ANNUAL INTERNAL RENEWABLE WATER RESOURCES[a]		ANNUAL WITHDRAWALS[b]			SECTORAL WITHDRAWALS (%)[c]		
	TOTAL (cubic km)	1995 PER CAPITA (cubic meters)	TOTAL (cubic km)	PERCENTAGE OF WATER RESOURCES	PER CAPITA (cubic meters)	DOMESTIC	INDUSTRY	AGRICULTURE
Tanzania	89.0	2,998	1.16	1	40	9	2	89
Togo	12.0	2,900	0.09	1	28	62	13	25
Tunisia	3.9	443	3.08	78	381	9	3	89
Uganda	66.0	3,099	0.20	0	20	32	8	60
Zaire	1,019.0	23,211	0.36	0	10	61	16	23
Zambia	116.0	12,267	1.71	1	186	16	7	77
Zimbabwe	20.0	1,776	1.22	6	136	14	7	79
NORTH AND MIDDLE AMERICA	**6,443.7**	**15,369**	**608.44**	**9**	**1,451**	**9**	**42**	**49**
Belize	16.0	74,419	0.02	0	109	10	0	90
Canada	2,901.0	98,462	45.10	2	1,602	18	70	12
Costa Rica	95.0	27,745	1.35	1	780	4	7	89
Cuba	34.5	3,125	8.10	23	870	9	2	89
Dominican Republic	20.0	2,557	2.97	15	446	5	6	89
El Salvador	19.0	3,285	1.00	5	245	7	4	89
Guatemala	116.0	10,922	0.73	1	139	9	17	74
Haiti	11.0	1,532	0.04	0	7	24	8	68
Honduras	63.4	11,216	1.52	2	294	4	5	91
Jamaica	8.3	3,392	0.32	4	159	7	7	86
Mexico	357.4	3,815	77.62	22	899	6	8	86
Nicaragua	175.0	39,477	0.89	1	367	25	21	54
Panama	144.0	54,732	1.30	1	754	12	11	77
Trinidad and Tobago	5.1	3,905	0.15	3	148	27	38	35
United States	2,478.0	9,413	467.34	19	1,870	13	45	42
SOUTH AMERICA	**9,526.0**	**29,788**	**106.21**	**1**	**332**	**18**	**23**	**59**
Argentina	994.0	28,739	27.60	4	1,043	9	18	73
Bolivia	300.0	40,464	1.24	0	201	10	5	85
Brazil	6,950.0	42,957	36.47	1	246	22	19	59
Chile	468.0	32,814	16.80	4	1,626	6	5	89
Colombia	1,070.0	30,483	5.34	0	174	41	16	43
Ecuador	314.0	27,400	5.56	2	581	7	3	90
Guyana	241.0	288,623	1.46	1	1,812	1	0	99
Paraguay	314.0	63,306	0.43	0	109	15	7	78
Peru	40.0	1,682	6.10	15	300	19	9	72
Suriname	200.0	472,813	0.46	0	1,189	6	5	89
Uruguay	124.0	38,920	0.65	1	241	6	3	91
Venezuela	1,317.0	60,291	4.10	0	382	43	11	46
ASIA	**13,206.7**	**3,819**	**1,633.85**	**12**	**542**	**6**	**9**	**85**
Afghanistan	50.0	2,482	26.11	52	1,830	1	0	99
Armenia	13.3	3,687	3.80	46	1,145	13	15	72
Azerbaijan	33.0	4,364	15.80	56	2,248	4	22	74
Bangladesh	2,357.0	19,571	22.50	1	220	3	1	96
Bhutan	95.0	57,998	0.02	0	14	36	10	54
Cambodia	498.1	48,590	0.52	0	64	5	1	94

(Continued on next page)

	ANNUAL INTERNAL RENEWABLE WATER RESOURCES[a]		ANNUAL WITHDRAWALS[b]			SECTORAL WITHDRAWALS (%)[c]		
	TOTAL (cubic km)	1995 PER CAPITA (cubic meters)	TOTAL (cubic km)	PERCENTAGE OF WATER RESOURCES	PER CAPITA (cubic meters)	DOMESTIC	INDUSTRY	AGRICULTURE
China	2,800.0	2,292	460.00	16	461	6	7	87
Georgia	65.2	11,942	4.00	7	741	21	37	42
India	2,085.0	2,228	380.00	18	612	3	4	93
Indonesia	2,530.0	12,804	16.59	1	96	13	11	76
Iran	117.5	1,746	45.40	39	1,362	4	9	87
Iraq	109.2	5,340	42.80	43	4,575	3	5	92
Israel	2.2	382	1.85	86	408	16	5	79
Japan	547.0	4,373	90.80	17	735	17	33	50
Jordan	1.7	314	0.45	32	173	29	6	65
Kazakhstan	169.4	9,900	37.90	30	2,294	4	17	79
Korea, North	67.0	2,801	14.16	21	687	11	16	73
Korea, South	66.1	1,469	27.60	42	632	19	35	46
Kuwait	0.2	103	0.50	X	525	64	32	4
Kyrgyzstan	61.7	13,003	11.70	24	2,729	3	7	90
Laos	270.0	55,305	0.99	0	259	8	10	82
Lebanon	5.6	1,854	0.75	16	271	11	4	85
Malaysia	456.0	22,642	9.42	2	768	23	30	47
Mongolia	24.6	10,207	0.55	2	273	11	27	62
Myanmar (Burma)	1,082.0	23,255	3.96	0	101	7	3	90
Nepal	170.0	7,756	2.68	2	150	4	1	95
Oman	1.9	892	0.48	24	564	3	3	94
Pakistan	468.0	3,331	153.40	33	2,053	1	1	98
Philippines	323.0	4,779	29.50	9	686	18	21	61
Saudi Arabia	4.6	254	3.60	164	497	45	8	47
Singapore	0.6	211	0.19	32	84	45	51	4
Sri Lanka	43.2	2,354	6.30	15	503	2	2	96
Syria	53.7	3,662	3.34	9	435	7	10	83
Tajikistan	101.3	16,604	12.6	13	2,455	5	7	88
Thailand	179.0	3,045	31.90	18	602	4	6	90
Turkey	193.1	3,117	33.50	17	585	24	19	57
Turkmenistan	72.0	17,573	22.80	33	6,390	1	8	91
United Arab Emirates	2.0	1,047	0.90	299	884	11	9	80
Uzbekistan	129.6	5,674	82.20	76	4,121	4	12	84
Vietnam	376.0	5,044	28.90	8	414	13	9	78
Yemen	5.2	359	3.40	136	335	5	2	93
EUROPE	**6,234.6**	**8,576**	**455.29**	**7**	**626**	**14**	**55**	**31**
Albania	21.3	6,190	0.20	1	94	6	18	76
Austria	90.3	11,333	2.36	3	304	33	58	9
Belarus	73.6	7,277	3.00	5	295	32	49	19
Belgium	12.5	1,236	9.03	72	917	11	85	4
Bosnia-Herzegovina	X	X	X	X	X	X	X	X
Bulgaria	205.0	23,378	13.90	7	1,544	3	76	22
Croatia	61.4	13,660	X	0	X	X	X	X
Czech Republic	58.2	5,653	2.74	5	266	41	57	2
Denmark	13.0	2,509	1.20	9	233	30	27	43
Estonia	17.6	11,490	3.30	21	2,097	5	92	3

	ANNUAL INTERNAL RENEWABLE WATER RESOURCES[a]		ANNUAL WITHDRAWALS[b]			SECTORAL WITHDRAWALS (%)[c]		
	TOTAL (cubic km)	1995 PER CAPITA (cubic meters)	TOTAL (cubic km)	PERCENTAGE OF WATER RESOURCES	PER CAPITA (cubic meters)	DOMESTIC	INDUSTRY	AGRICULTURE
Finland	113.0	22,126	2.20	2	440	12	85	3
France	198.0	3,415	37.73	19	665	16	69	15
Germany	171.0	2,096	46.27	27	579	11	70	20
Greece	58.7	5,612	5.04	9	523	8	29	63
Hungary	120.0	11,864	6.81	6	661	9	55	36
Iceland	168.0	624,535	0.16	0	636	31	63	6
Ireland	50.0	14,073	0.79	2	233	16	74	10
Italy	167.0	2,920	56.20	34	986	14	27	59
Latvia	34.0	13,297	0.70	2	262	42	44	14
Lithuania	24.2	6,541	4.40	19	1,190	7	90	3
Macedonia	X	X	X	X	X	X	X	X
Moldova	13.7	3,093	3.70	29	853	7	70	23
Netherlands	90.0	5,805	7.81	9	518	5	61	34
Norway	392.0	90,385	2.03	1	488	20	72	8
Poland	56.2	1,464	12.28	22	321	13	76	11
Portugal	69.6	7,085	7.29	10	739	15	37	48
Romania	208.0	9,109	26.00	13	1,134	8	33	59
Russia	4,498.0	30,599	117.00	3	790	17	60	23
Slovakia	30.8	5,753	1.78	6	337	X	X	X
Slovenia	X	X	X	X	X	X	X	X
Spain	111.3	2,809	30.75	28	781	12	26	62
Sweden	180.0	20,501	2.93	2	341	36	55	9
Switzerland	50.0	6,943	1.19	2	173	23	73	4
Ukraine	231.0	4,496	34.70	40	673	16	54	30
United Kingdom	71.0	1,219	11.79	17	205	20	77	3
Yugoslavia	X	X	X	X	X	X	X	X
OCEANIA	**1,614.3**	**56,543**	**16.73**	**1**	**586**	**64**	**2**	**34**
Australia	343.0	18,963	14.60	4	933	65	2	33
Fiji	28.6	36,416	0.03	0	42	20	20	60
New Zealand	327.0	91,469	2.00	1	589	46	10	44
Papua New Guinea	801.0	186,192	0.10	0	28	29	22	49
Solomon Islands	44.7	118,254	0.00	0	0	40	20	40

[a]Annual renewable water resources usually include river flows from other countries.

[b]Withdrawal data from most recent year available; varies by country from 1987 to 1995.

[c]Total withdrawals may exceed 100% because of groundwater withdrawals or river inflows.

Source: World Resources 1996–97.

Table G
Access to Safe Drinking Water and Sanitation Services, 1980–1990

	PERCENT OF POPULATION WITH ACCESS TO SAFE DRINKING WATER				PERCENT OF POPULATION WITH ACCESS TO SANITATION SERVICES			
	URBAN		RURAL		URBAN		RURAL	
	1980	1990	1980	1990	1980	1990	1980	1990
AFRICA								
Algeria	X	X	X	X	X	X	X	X
Angola	85	73	10	20	40	25	15	20
Benin	26	73	15	43	48	60	4	35
Botswana	X	100	X	88	X	100	X	85
Burkina Faso	27	44	31	70	38	35	5	5
Burundi	90	92	20	43	40	64	35	16
Cameroon	X	42	X	45	X	X	X	X
Central African Republic	X	19	X	26	X	45	X	46
Chad	X	X	X	X	X	X	X	X
Congo	36	92	3	2	17	X	0	2
Djibouti	50	50	20	21	43	94	20	50
Egypt	88	95	64	86	45	80	10	26
Equatorial Guinea	47	65	X	18	99	54	X	24
Eritrea	X	X	X	X	X	X	X	X
Ethiopia	X	70	X	11	X	97	X	7
Gabon	X	90	X	50	X	X	X	X
Gambia	85	100	X	48	X	100	X	27
Ghana	72	63	33	39	47	63	17	60
Guinea	69	100	2	37	54	65	1	0
Guinea-Bissau	18	18	8	27	21	30	13	18
Ivory Coast	X	57	X	80	X	81	X	100
Kenya	85	X	15	X	89	X	19	X
Lesotho	37	59	11	45	13	14	14	23
Liberia	X	93	16	22	18	4	5	8
Libya	100	100	90	80	100	100	72	85
Madagascar	80	62	7	10	9	X	X	X
Malawi	77	66	37	49	100	X	81	X
Mali	37	41	0	4	79	81	0	10
Mauritania	80	67	85	65	5	34	X	X
Mauritius	100	100	98	100	100	100	90	100
Morocco	100	100	X	18	X	100	X	19
Mozambique	X	44	X	17	X	61	X	11
Namibia	X	90	X	37	X	24	X	11
Niger	41	98	32	45	36	71	3	4
Nigeria	60	100	30	22	X	80	X	11
Rwanda	48	84	55	67	60	88	50	17
Senegal	33	65	25	26	5	57	2	38
Sierra Leone	50	80	2	20	31	55	6	31
Somalia	60	50	20	29	45	41	5	5
South Africa	X	X	X	X	X	X	X	X
Sudan	X	90	31	20	63	40	0	5
Swaziland	X	100	X	7	X	100	X	25
Tanzania	X	75	X	46	X	76	X	77
Togo	70	100	31	61	24	42	10	16
Tunisia	100	100	17	31	100	71	X	15
Uganda	45	60	8	30	40	32	10	60
Zaire	X	68	X	24	X	46	X	11
Zambia	65	76	32	43	100	77	48	34
Zimbabwe	X	95	X	80	X	95	X	22
NORTH AND MIDDLE AMERICA								
Belize	X	94	36	53	62	76	75	22
Canada	X	100	X	100	X	X	X	X
Costa Rica	100	100	68	84	93	100	82	93
Cuba	X	100	X	91	X	100	X	68
Dominican Republic	85	82	33	45	25	95	4	75
El Salvador	67	87	40	15	80	85	26	38
Guatemala	89	92	18	43	45	72	20	52
Haiti	48	56	8	35	39	44	10	17

(Continued on next page)

	PERCENT OF POPULATION WITH ACCESS TO SAFE DRINKING WATER				PERCENT OF POPULATION WITH ACCESS TO SANITATION SERVICES			
	URBAN		RURAL		URBAN		RURAL	
	1980	1990	1980	1990	1980	1990	1980	1990
Honduras	50	85	40	48	40	89	26	42
Jamaica	X	95	X	46	X	14	X	X
Mexico	64	94	43	49	51	85	12	12
Nicaragua	91	76	10	21	35	32	X	X
Panama	100	100	65	66	62	100	28	68
Trinidad and Tobago	100	100	93	88	95	100	88	92
United States	X	X	X	X	X	X	X	X
SOUTH AMERICA								
Argentina	65	73	17	17	89	100	32	29
Bolivia	69	76	10	30	37	38	4	14
Brazil	80	95	51	61	32	84	X	32
Chile	100	100	17	21	99	100	X	6
Colombia	X	87	79	82	100	84	4	18
Ecuador	82	63	16	44	39	56	14	38
Guyana	X	100	60	71	100	97	80	81
Paraguay	39	61	10	9	95	31	89	60
Peru	68	68	21	24	57	76	0	20
Suriname	X	82	79	56	100	64	79	36
Uruguay	96	100	2	5	59	60	60	65
Venezuela	91	89	50	36	90	97	70	72
ASIA								
Afghanistan	28	40	8	19	X	13	X	X
Armenia	X	X	X	X	X	X	X	X
Azerbaijan	X	X	X	X	X	X	X	X
Bangladesh	26	39	40	89	21	40	1	4
Bhutan	50	60	5	30	X	80	X	3
Cambodia	X	X	X	X	X	X	X	X
China	X	87	X	68	X	100	X	81
Georgia	X	X	X	X	X	X	X	X
India	77	86	31	69	27	44	1	3
Indonesia	35	35	19	33	29	79	21	30
Iran	X	100	X	75	X	100	X	35
Iraq	X	93	X	41	X	96	X	18
Israel	X	100	X	97	X	99	X	95
Japan	X	100	X	85	X	X	X	X
Jordan	100	100	65	97	94	100	34	100
Kazakhstan	X	X	X	X	X	X	X	X
Korea, North	X	100	X	100	X	100	X	100
Korea, South	86	100	61	76	67	100	12	100
Kuwait	X	100	X	X	X	100	X	X
Kyrgyzstan	X	X	X	X	X	X	X	X
Laos	21	47	12	25	11	30	3	8
Lebanon	100	X	100	X	94	X	18	X
Malaysia	90	96	49	66	100	94	55	94
Mongolia	X	100	X	58	X	100	X	47
Myanmar	38	79	15	72	38	50	15	13
Nepal	83	66	7	34	16	34	1	3
Oman	X	87	X	42	X	100	X	34
Pakistan	72	82	20	42	42	53	2	12
Philippines	65	93	43	72	81	79	67	63
Saudi Arabia	92	100	87	74	81	100	50	30
Singapore	100	100	X	X	80	99	X	2
Sri Lanka	65	80	18	55	80	68	63	45
Syria	98	91	54	68	74	72	28	55
Tajikistan	X	X	X	X	X	X	X	X
Thailand	65	67	63	85	64	84	41	86
Turkey	95	100	62	70	56	95	X	90
Turkmenistan	X	X	X	X	X	X	X	X
United Arab Emirates	95	100	81	100	93	100	22	77
Uzbekistan	X	X	X	X	X	X	X	X
Vietnam	X	47	32	33	X	23	55	10
Yemen	X	X	X	X	X	X	X	X

(Continued on next page)

	PERCENT OF POPULATION WITH ACCESS TO SAFE DRINKING WATER				PERCENT OF POPULATION WITH ACCESS TO SANITATION SERVICES			
	URBAN		RURAL		URBAN		RURAL	
	1980	1990	1980	1990	1980	1990	1980	1990
EUROPE								
Albania	X	100	X	95	X	100	X	100
Austria	X	100	X	100	X	100	X	100
Belarus	X	100	X	100	X	100	X	100
Belgium	X	100	X	100	X	100	X	100
Bulgaria	X	100	X	96	X	100	X	100
Czech Republic	X	100	X	100	X	100	X	100
Denmark	X	100	X	100	X	100	X	100
Estonia	X	X	X	X	X	X	X	X
Finland	X	99	X	90	X	100	X	100
France	X	100	X	100	X	100	X	100
Germany	X	100	X	100	X	X	X	X
Greece	X	100	X	95	X	100	X	95
Hungary	X	100	X	95	X	100	X	100
Iceland	X	100	X	100	X	100	X	100
Ireland	X	100	X	100	X	100	X	100
Italy	X	100	X	100	X	100	X	100
Latvia	X	X	X	X	X	X	X	X
Lithuania	X	X	X	X	X	X	X	X
Moldova	X	X	X	X	X	X	X	X
Netherlands	X	100	X	100	X	100	X	100
Norway	X	100	X	100	X	100	X	100
Poland	X	94	X	82	X	100	X	100
Portugal	X	97	X	90	X	100	X	95
Romania	X	100	X	90	X	100	X	95
Russia	X	X	X	X	X	X	X	X
Slovakia	X	X	X	X	X	X	X	X
Spain	X	100	X	100	X	100	X	100
Sweden	X	100	X	100	X	100	X	100
Switzerland	X	100	X	100	X	100	X	100
Ukraine	X	X	X	X	X	X	X	X
United Kingdom	X	100	X	100	X	100	X	100
Yugoslavia (former)[1]	X	100	X	65	X	78	X	46
OCEANIA								
Australia	X	100	X	100	X	100	X	100
Fiji	94	96	66	69	85	91	60	65
New Zealand	X	100	X	82	X	100	X	88
Papua New Guinea	55	94	10	20	96	57	3	56
Solomon Islands	91	82	20	58	82	73	10	2

[1] In this and subsequent tables, figures for former Yugoslavia encompass the current states of Bosnia-Herzegovina, Croatia, Macedonia, Slovenia, and Yugoslavia (Serbia-Montenegro).

Source: World Resources 1994–95.

Table H

Marine and Freshwater Catches, Aquaculture, and Fish Consumption

	AVERAGE ANNUAL MARINE CATCH (thousands of metric tons)	AVERAGE ANNUAL FRESHWATER CATCH (thousands of metric tons)	AVERAGE ANNUAL AQUACULTURE PRODUCTION: FISH AND SHELLFISH (thousands of metric tons)	PER CAPITA ANNUAL FOOD SUPPLY FROM FISH AND SEAFOOD (kg)	PERCENT CHANGE IN ANNUAL FOOD SUPPLY FROM FISH AND SEAFOOD (1980–1990)
AFRICA	**3,180.71**	**1,837.3**	**67.2**	**7.9**	**−2.9**
Algeria	89.88	0.4	0.3	4.8	114.9
Angola	90.05	7.7	X	23.3	99.1
Benin	8.94	29.5	0.1	9.6	−26.5
Botswana	X	1.9	X	2.9	0.0
Burkina Faso	X	7.3	0.0	2.1	52.4
Burundi	X	17.4	0.0	2.2	−37.4
Cameroon	57.10	20.7	0.2	13.4	18.2
Central African Republic	X	13.2	0.1	4.8	−19.0
Chad	X	61.7	X	17.1	−4.5
Congo	20.68	25.9	0.2	34.8	22.5
Djibouti	0.38	0.0	X	3.3	312.5
Egypt	78.90	199.5	39.4	7.2	49.3
Equatorial Guinea	3.35	0.4	X	X	X
Eritrea	X	X	X	X	X
Ethiopia	1.77	2.8	0.0	0.1	0.0
Gabon	20.50	2.0	0.0	28.6	−24.9
Gambia	17.82	2.6	0.1	13.7	42.2
Ghana	315.02	57.6	0.4	24.9	−3.4
Guinea	32.33	3.2	0.0	7.9	29.5
Guinea-Bissau	5.07	0.2	X	3.5	0.0
Ivory Coast	68.88	27.4	0.2	15.7	−13.6
Kenya	8.31	174.0	1.1	5.5	80.2
Lesotho	X	0.0	0.0	1.5	64.3
Liberia	6.30	4.0	0.0	13.8	−11.2
Libya	7.81	0.0	0.1	3.5	−50.2
Madagascar	71.08	30.0	0.2	8.0	27.0
Malawi	X	69.5	0.2	10.4	2.3
Mali	X	65.6	0.0	7.4	−34.5
Mauritania	85.20	6.0	X	8.6	−21.6
Mauritius	16.85	0.1	0.1	17.8	1.3
Morocco	558.02	1.6	0.2	7.8	23.2
Mozambique	33.77	0.3	0.0	3.0	−17.4
Namibia	160.07	0.1	X	9.0	18.4
Niger	X	3.8	0.1	0.4	−66.7
Nigeria	192.68	101.5	16.6	6.5	−48.4
Rwanda	X	2.5	0.1	0.2	−25.0
Senegal	282.90	17.2	0.0	21.3	−6.6
Sierra Leone	36.59	15.3	0.0	13.6	−36.4
Somalia	16.86	0.4	X	2.2	37.5
South Africa	635.66	2.3	4.0	10.1	7.9
Sudan	1.40	30.4	0.2	0.9	−36.4
Swaziland	X	0.1	0.0	0.1	0.0
Tanzania	53.86	343.3	0.4	14.9	31.4
Togo	10.14	0.4	0.0	15.0	36.5
Tunisia	94.51	0.0	1.0	10.8	28.7
Uganda	X	237.4	0.0	13.3	−11.3
Zaire	2.00	160.7	0.7	8.0	22.3
Zambia	X	65.7	1.1	8.2	−14.5
Zimbabwe	X	24.0	0.2	1.9	−3.3
NORTH AND MIDDLE AMERICA	**8,820.01**	**536.1**	**474.6**	**17.4**	**27.7**
Belize	1.64	0.0	0.3	6.5	−6.2
Canada	1,525.48	50.2	33.3	25.1	20.1
Costa Rica	16.8	1.0	1.0	4.0	−49.8
Cuba	160.60	21.1	21.0	20.0	24.9
Dominican Republic	18.07	1.6	0.4	4.7	−39.0

(Continued on next page)

	AVERAGE ANNUAL MARINE CATCH (thousands of metric tons)	AVERAGE ANNUAL FRESHWATER CATCH (thousands of metric tons)	AVERAGE ANNUAL AQUACULTURE PRODUCTION: FISH AND SHELLFISH (thousands of metric tons)	PER CAPITA ANNUAL FOOD SUPPLY FROM FISH AND SEAFOOD (kg)	PERCENT CHANGE IN ANNUAL FOOD SUPPLY FROM FISH AND SEAFOOD (1980–1990)
El Salvador	6.69	4.0	0.5	1.8	−11.7
Guatemala	3.60	2.1	0.9	0.5	−37.5
Haiti	4.93	0.3	X	4.0	26.3
Honduras	17.63	0.2	3.6	3.3	120.0
Jamaica	7.23	3.3	2.5	18.7	−5.9
Mexico	1,253.83	179.5	56.7	10.1	24.1
Nicaragua	4.29	0.2	0.1	0.7	−33.3
Panama	162.07	0.3	3.7	13.4	−11.3
Trinidad and Tobago	8.42	0.0	X	7.4	−42.1
United States	5,425.79	272.1	350.2	21.3	36.3
SOUTH AMERICA	**14,958.25**	**332.4**	**155.5**	**8.2**	**−7.9Z**
Argentina	550.16	10.8	0.4	6.3	−8.3
Bolivia	X	3.6	0.4	0.9	−63.9
Brazil	606.25	211.4	21.0	6.0	−12.5
Chile	5,879.99	4.2	31.9	22.9	11.3
Colombia	77.59	34.1	9.5	2.6	−33.6
Ecuador	502.87	2.0	86.0	10.8	−8.0
Guyana	36.86	0.8	0.1	43.0	7.7
Paraguay	X	12.2	0.1	2.4	132.3
Peru	6,858.88	32.1	5.7	23.2	−9.2
Suriname	3.79	0.1	0.0	6.6	−65.5
Uruguay	118.43	0.3	X	3.5	−49.5
Venezuela	312.65	20.8	0.9	13.5	17.1
ASIA	**35,167.85**	**10,283.3**	**9,697.4**	**11.7**	**23.6**
Afghanistan	X	1.5	X	0.1	0.0
Armenia	X	X	X	X	X
Azerbaijan	X	X	X	X	X
Bangladesh	254.63	606.7	170.1	7.2	−4.0
Bhutan	X	1.0	0.0	X	X
Cambodia	34.12	63.4	6.2	8.7	38.1
China	6,942.47	5,207.6	5,481.8	9.6	100.0
Georgia	X	X	X	X	X
India	2,271.78	1,551.9	1,078.1	3.3	7.5
Indonesia	2,272.05	779.8	482.2	13.9	23.1
Iran	201.73	65.8	32.1	4.6	183.7
Iraq	3.83	10.9	5.0	1.0	−71.2
Israel	6.77	15.9	14.8	21.8	28.7
Japan	10,072.51	203.2	787.2	71.7	11.5
Jordan	0.00	0.0	0.0	2.5	−13.8
Kazakhstan	X	X	X	X	X
Korea, North	1,613.43	103.3	74.5	42.6	32.6
Korea, South	2,695.71	31.8	345.8	47.6	21.2
Kuwait	4.71	0.0	0.0	11.8	29.6
Kyrgyzstan	X	X	X	X	X
Laos	X	20.0	3.4	5.0	−21.2
Lebanon	1.61	0.1	0.1	0.7	5.0
Malaysia	596.27	14.9	54.3	27.5	−36.8
Mongolia	X	0.2	X	1.1	−15.0
Myanmar	594.62	154.3	6.1	15.3	5.0
Nepal	X	14.2	8.8	0.7	250.0
Oman	118.57	0.0	X	X	X
Pakistan	368.90	111.4	10.7	1.9	7.7
Philippines	1,622.53	579.0	385.0	33.5	6.5
Saudi Arabia	46.08	1.5	1.6	7.5	−3.8
Singapore	12.94	0.1	1.9	28.4	−2.1
Sri Lanka	157.98	31.6	5.2	14.6	10.3
Syria	1.56	3.9	2.5	0.5	−75.8
Tajikistan	X	X	X	X	X
Thailand	2,613.91	236.6	289.3	19.6	3.5
Turkey	357.12	32.2	6.0	6.7	12.4

(Continued on next page)

	AVERAGE ANNUAL MARINE CATCH (thousands of metric tons)	AVERAGE ANNUAL FRESHWATER CATCH (thousands of metric tons)	AVERAGE ANNUAL AQUACULTURE PRODUCTION: FISH AND SHELLFISH (thousands of metric tons)	PER CAPITA ANNUAL FOOD SUPPLY FROM FISH AND SEAFOOD (kg)	PERCENT CHANGE IN ANNUAL FOOD SUPPLY FROM FISH AND SEAFOOD (1980–1990)
Turkmenistan	X	X	X	X	X
United Arab Emirates	92.86	0.0	0.0	24.4	27.9
Uzbekistan	X	X	X	X	X
Vietnam	607.67	257.3	153.7	12.6	14.8
Yemen	78.22	0.7	X	5.3	−35.5
EUROPE	**11,263.84**	**459.2**	**1,193.0**	**18.8**	**23.1**
Albania	6.61	4.9	2.5	3.0	−12.6
Austria	X	4.8	4.2	8.7	56.0
Belarus	X	X	X	X	X
Belgium	39.82	0.8	0.7	19.8	4.9
Bulgaria	59.63	9.7	9.1	6.2	−11.0
Czech Republic	X	22.2	21.9	6.9	40.8
Denmark	1,712.71	33.1	39.0	19.1	0.5
Estonia	X	X	X	X	X
Finland	86.31	8.5	18.8	31.2	7.6
France	828.15	45.2	263.3	30.6	28.4
Germany	313.17	51.7	67.4	12.0	12.5
Greece	133.67	10.0	8.2	18.2	14.5
Hungary	X	32.9	17.5	4.8	26.3
Iceland	1,354.38	0.7	2.5	93.0	9.4
Ireland	219.14	0.7	25.1	16.4	18.6
Italy	480.59	55.0	137.6	20.1	44.4
Latvia	X	X	X	X	X
Lithuania	X	X	X	X	X
Moldova	X	X	X	X	X
Netherlands	447.56	3.7	87.3	8.0	−23.2
Norway	1,904.84	0.5	119.6	44.2	4.7
Poland	452.57	42.4	27.3	13.4	8.9
Portugal	324.19	2.1	6.7	57.8	111.7
Romania	107.29	51.8	34.3	8.2	18.3
Russia	9,290.91	952.4	392.6	29.0	13.6
Slovakia	48.56	20.1	18.7	7.5	31.2
Spain	1,423.61	29.7	215.3	37.7	20.6
Sweden	248.04	5.7	8.4	27.6	−1.7
Switzerland	X	4.5	1.1	13.2	47.2
Ukraine	X	X	X	X	X
United Kingdom	802.58	18.2	52.9	18.7	16.1
Yugoslavia (former)	37.16	20.4	9.3	3.2	3.3
OCEANIA	**938.19**	**21.8**	**49.3**	**21.6**	**32.2**
Australia	205.72	4.1	13.6	16.8	28.9
Fiji	28.76	4.2	0.0	40.4	26.0
New Zealand	577.80	1.1	34.8	36.8	71.2
Papua New Guinea	11.79	12.3	0.0	22.0	21.5
Solomon Islands	60.36	0.0	0.0	56.8	−7.4

Source: World Resources 1994–95.

Part V

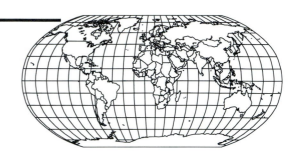

Human Impact on the Biosphere

Map 34 Cropland Per Capita: Changes, 1983–1993

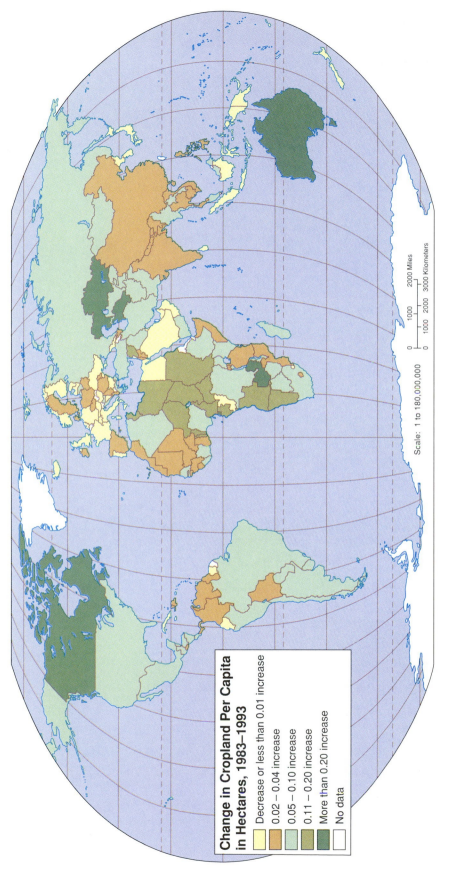

Change in Cropland Per Capita in Hectares, 1983–1993

- Decrease or less than 0.01 increase
- 0.02 – 0.04 increase
- 0.05 – 0.10 increase
- 0.11 – 0.20 increase
- More than 0.20 increase
- No data

Scale: 1 to 180,000,000

0   1000   2000 Miles

0   1000   2000   3000 Kilometers

As population has increased rapidly throughout the world, area in cultivated land has increased at the same time; in fact, the amount of farmland per person has gone up slightly. Unfortunately, the figures that show this also tell us that since most of the best (or even good) agricultural land in 1983 was already under cultivation, most of the agricultural area added since the early 1980s involves land that would have been viewed as marginal by the fathers and grandfathers of present farmers—marginal in that it was too dry, too wet, too steep to cultivate, too far from a market, and so on. The continued expansion of agricultural area is one reason that serious famine and starvation have struck only a few regions of the globe. But land, more than any other resource we deal with, is finite, and the expansion cannot continue indefinitely.

Map $\boxed{35}$ Annual Change in Forest Cover, 1981–1990

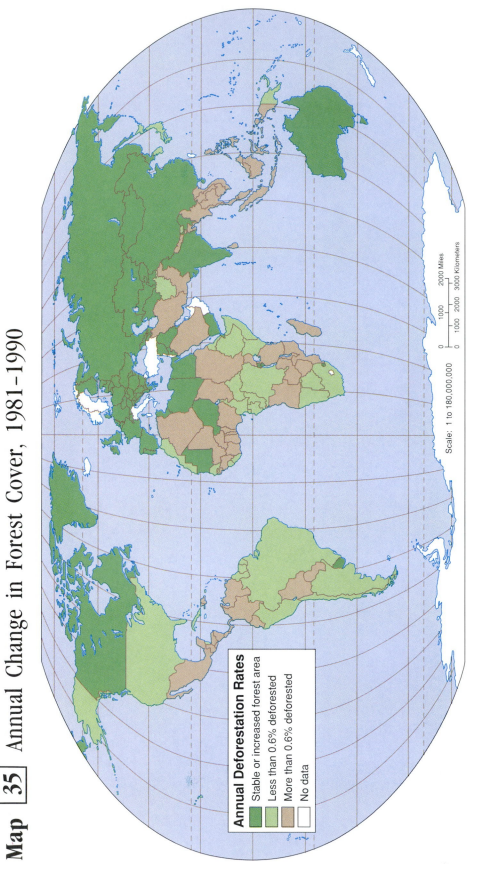

Annual Deforestation Rates

- Stable or increased forest area
- Less than 0.6% deforested
- More than 0.6% deforested
- No data

Scale: 1 to 180,000,000

0   1000   2000 Miles

0   1000   2000   3000 Kilometers

One of the most discussed environmental problems is that of deforestation. For most people, deforestation means clearing of tropical rain forests for agricultural purposes. Yet nearly as much forest land per year—much of it in North America, Europe, and Russia—is impacted by commercial lumbering as is cleared by tropical farmers and ranchers. Even in the tropics, much of the forest clearance is undertaken by large corporations producing high-value tropical hardwoods for the global market in furniture, ornaments, and other fine wood products. Still, it is the agriculturally driven clearing of the great rain forests of the Amazon Basin, West and central Africa, Middle America, and Southeast Asia that draws public attention. Although much concern over forest clearance focuses on the relationship between forest clearance and the reduction in the capacity of the world's vegetation system to absorb carbon dioxide (and thus delay global warming), of just as great concern are issues having to do with the loss of biodiversity (large numbers of plants and animals), the near-total destruction of soil systems, and disruptions in water supply that accompany clearing.

Map 36 Forest Defoliation: The European "Dead Zone"

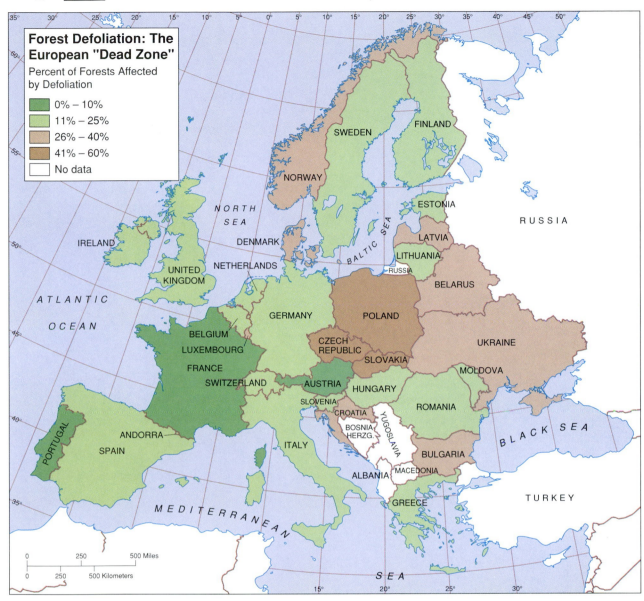

Forest Defoliation: The European "Dead Zone"

Percent of Forests Affected by Defoliation

- 0% – 10%
- 11% – 25%
- 26% – 40%
- 41% – 60%
- No data

While world attention tends to focus on the problems of the tropical forests, a different forest problem has emerged in the much different environment of eastern and central Europe. Years of minimal environmental protection, coupled with the industrial demands of centrally planned economies and with cheap, locally available high-sulfur coal, have produced what is quite possibly one of the world's most ravaged regions. The most visible signs of the pollution that rages across the countries of the former Warsaw Pact are the forest "dead zones"—regions of tens of thousands of acres of forestland that have been destroyed by various forms of air and water pollution. The major destroyer of forests is acid precipitation (rain or snow) resulting from the combination of the hydrogen sulfide emissions from factories and power-generating plants with atmospheric water vapor. Rainfall with an acidity level similar to that of lemon juice produces bark and leaf damage that inhibits the ability of vegetation to photosynthesize carbohydrates. The resulting diminished food supply weakens and eventually kills trees and other plants. Much of eastern Europe's once-mighty forests are now stands of charred timber, stumps and trunks blackened not by fire but by the chemicals generated by human industry.

Map 37 Clearing the Tropical Forests: The Amazon Basin and Rondônia

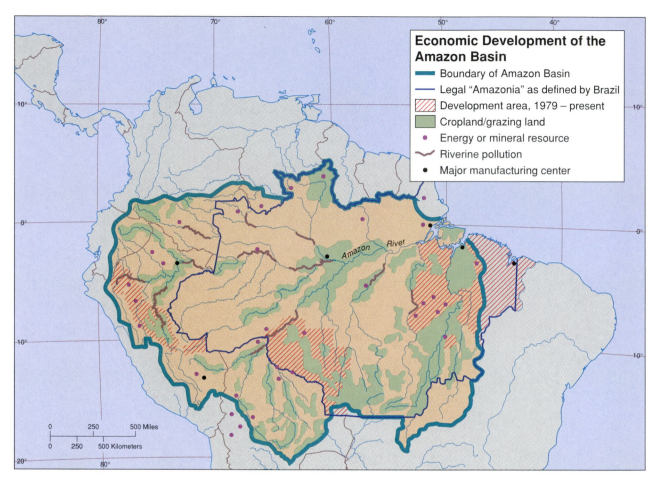

Economic Development of the Amazon Basin

- Boundary of Amazon Basin
- Legal "Amazonia" as defined by Brazil
- Development area, 1979 – present
- Cropland/grazing land
- Energy or mineral resource
- Riverine pollution
- Major manufacturing center

The most rapid and destructive tropical forest clearance is currently taking place in the drainage basin of the Amazon River in South America, especially in the region designated as "Amazonia" by the Brazilian government. The clearance is most notable in those areas of Amazonia designated as "development areas," such as Rondônia state. Reasons for and consequences of forest clearance in this region are numerous. Mineral wealth in gold, copper, manganese, iron, tin, nickel, coal, diamonds, and oil attracts mining developments that rip up forests to get at the rich minerals below, while the extraction of gold and other metals despoils rivers with toxins used in processing. Open space and cheap land attract farmers, but the fragile tropical soil is quickly exhausted and areas are left barren, the depleted soil washing into streams and rivers and adding to already enormous sedimentary loads. In addition, forest products valuable to the furniture industry tempt lumber companies to destroy hundreds of acres for the sake of a few trees. Finally, the cultivation of coca and production of cocaine in the western Amazon Basin leave rivers polluted with the chemicals used in refining the drug. All of these processes come together in Rondônia. Although a number of Indian reserves, national forests, and national parks are in the state, farmers, loggers, and miners encroach upon the protected or reserved lands without interference, since there are too few government officials to patrol the land effectively. Brazil has designed a number of "agroecological zones" for Rondônia, intended to allow only sustainable use of the region's resources, in an attempt to defend the forest from unscrupulous developers. Unfortunately, with

the exception of the extensive shifting cultivation agriculture of the indigenous peoples, Rondônia and the rest of the Amazon Basin simply do not lend themselves to modern uses. In a typical cycle, crop farmers clear patches of land and harvest a few years' worth of crops before the soil becomes too unproductive to support anything but pasture grasses. Farms are abandoned in favor of grasslands for cattle and sheep grazing; within a decade these pastures, too, become exhausted and little is left but bare soil. In a process known as "laterization," this soil often converts to concentrates of iron and aluminum oxides that are as dense as bedrock. When this happens, the forest may never be able to re-cover the area.

Map 38 Global Production of Wood

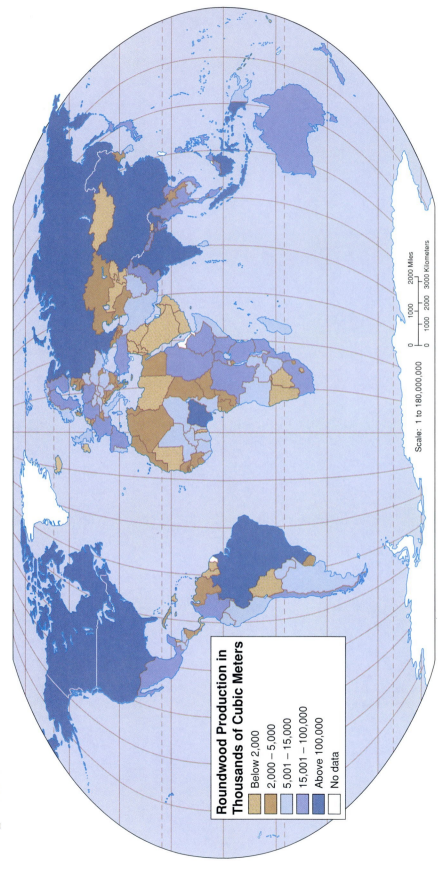

Roundwood Production in Thousands of Cubic Meters

- Below 2,000
- 2,000 – 5,000
- 5,001 – 15,000
- 15,001 – 100,000
- Above 100,000
- No data

Scale: 1 to 180,000,000

0   1000   2000 Miles

0   1000   2000   3000 Kilometers

The production of wood for fuel and charcoal and for manufacturing and construction purposes is a measure of a country's willingness to harvest natural resources, sometimes in a sustainable fashion and sometimes not. It is also a measure, obviously, of the amount of forest land available in a country, although the degree to which even desert countries produce wood (mostly for fuel and charcoal purposes) is surprising. Wood production is important as an environmental indicator because it portrays the tendency of environmental systems to be changed by one of the oldest resource extractive activities, forest clearance. The map indicates clearly that it is not just the tropical forests of the world that are threatened by removal. Just as threatened are the midlatitude and high-latitude forests of the more developed parts of North America, Europe, and Asia.

Map 39 The Loss of Biodiversity: Globally Threatened Animal Species

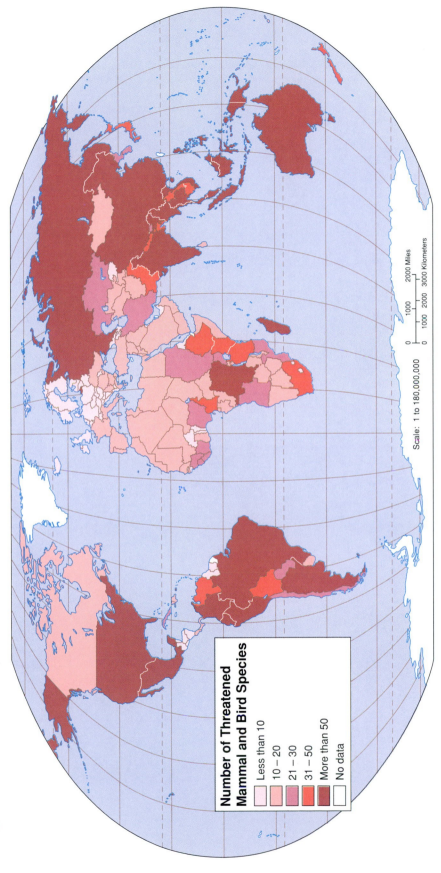

Number of Threatened Mammal and Bird Species

Less than 10
10 – 20
21 – 30
31 – 50
More than 50
No data

Scale: 1 to 180,000,000

0 1000 2000 2000 Miles
0 1000 2000 3000 Kilometers

Threatened species are those in grave danger of going extinct. Their populations are becoming restricted in range, and the size of the populations required for sustained breeding is nearing a critical minimum. This classification is distinct from endangered species, which are in immediate danger of becoming extinct. Their range is already so reduced that the animals may no longer be able to move freely within an ecozone, and their populations are at the level where the species may no longer be able to sustain breeding. Most species become threatened first and then endangered as their range and numbers continue to decrease. When people think of animal extinction, they think of large herbivorous species like the rhinoceros or fierce carnivores like lions, tigers, or grizzly bears. Certainly these animals make almost any list of endangered or threatened species. But for every rhino or elephant, there are literally hundreds of less conspicuous animals

that are equally threatened. Barring cataclysmic events like a meteor strike, extinction is normally nature's way of informing a species that it is inefficient. Indeed, it is usually the larger animals, with long gestation periods, low reproduction rates, and high resource demands, that tend to disappear most frequently. But conditions in the late twentieth century are controlled more by human activities than by natural evolutionary processes. Species that are endangered or threatened fall into that category because, somehow, they are competing with us or with our domesticated livestock for space and food. Hunting (poaching) for the ivory tusks of the elephant or the horn of the rhino is dramatic and tragic. But poaching had less to do with those species' predicament than the simple fact that rhinos and elephants compete with humans for food and space. And in that competition the animals are always going to lose.

– 85 –

Map 40 The Loss of Biodiversity: Globally Threatened Plant Species

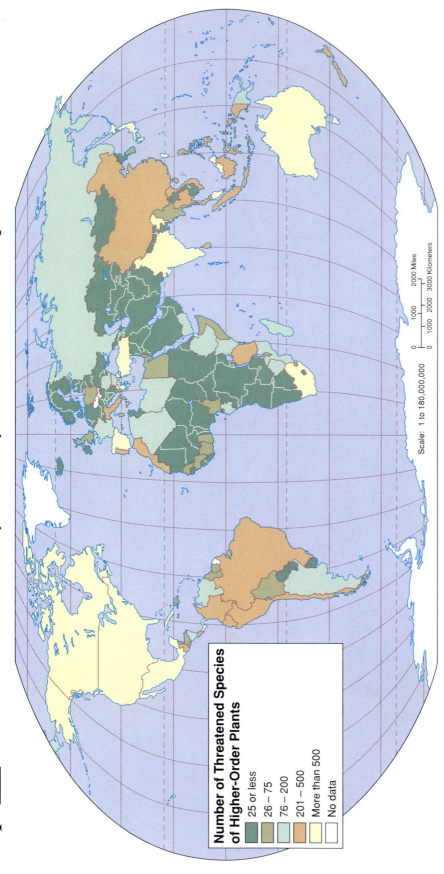

Number of Threatened Species of Higher-Order Plants

- 25 or less
- 26 – 75
- 76 – 200
- 201 – 500
- More than 500
- No data

Scale: 1 to 180,000,000

0   1000   2000 Miles

0   1000   2000   3000 Kilometers

While most people tend to be more concerned about the animals on threatened and endangered species lists, the fact is that many more plants are in jeopardy, and the loss of plant life is, in all ecological regions, a more critical occurrence than the loss of animal populations. Plants are the primary producers in the ecosystem; that is, plants produce the food upon which all other species in the food web, including human beings, depend for sustenance. It is plants from which many of our critical medicines come, and it is plants that maintain the delicate balance between soil and water in most of the world's regions. When environmental scientists speak of a

loss of "biodiversity," what they are most often describing is a loss of the richness and complexity of plant life that lends stability to ecosystems. Systems with more plant life tend to be more stable than those with less. For these and other reasons, the scientific concern over extinction is greater when applied to plants than to animals. It is difficult for people to become as emotional over a teak tree as they would over an elephant. But as great a tragedy as the loss of the elephant would be, the loss of the teak tree would be greater.

Map **41** The World's Wilderness Areas, 1500 and 2000

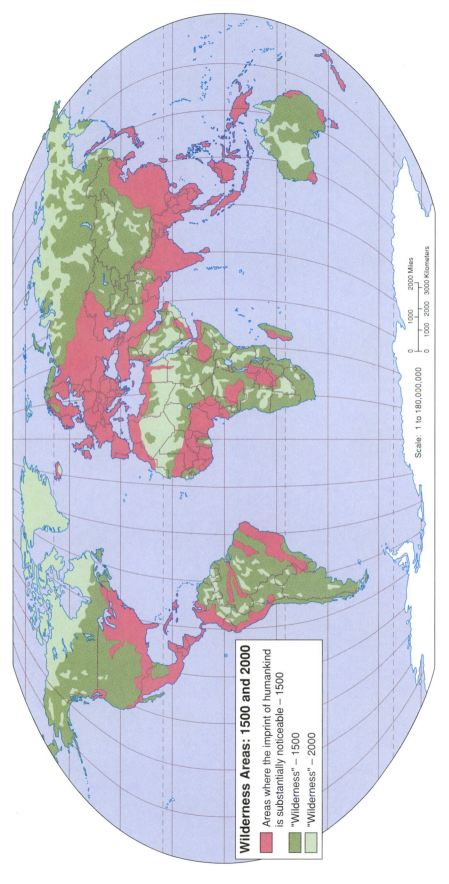

Wilderness Areas: 1500 and 2000

- ■ Areas where the imprint of humankind is substantially noticeable – 1500
- ■ "Wilderness" – 1500
- ■ "Wilderness" – 2000

Scale: 1 to 180,000,000

2000 Miles
1000

0   1000   2000   3000 Kilometers

The U.S. Wilderness Act of 1964 defined "wilderness" as land that "generally appears to have been affected primarily by the forces of nature, with the imprint of man's work substantially unnoticeable." The map portrays what was probably wilderness according to this definition in the year 1500, before the major impact of the global European expansion was felt, and in the year 2000, after 500 years of economic growth and environmental exploitation. Most of the area that could have been defined as "wilderness" in 1500 would now be defined as land bearing the significant imprint of human activities. By 2000, precious little of the earth's surface will be in a natural state. Whether or not this is a problem is largely a philosophical issue rather than a scientific one as human-impacted or controlled ecosystems have the potential to be as productive and stable as purely natural ones. For many people (and for most Europeans) in the year 1500 the "wilderness" was something to be subdued and controlled and made productive. Wilderness was not seen as something of value—quite the contrary. In contemporary terms, wilderness has different connotations, and we now tend to see value in preserving areas in their natural state "with the imprint of man's work substantially unnoticeable." Whether we will be able to do this in the face of increasing populations and resource demand is highly questionable.

Map 42 Vegetation Zones of the Greenhouse World, 2050

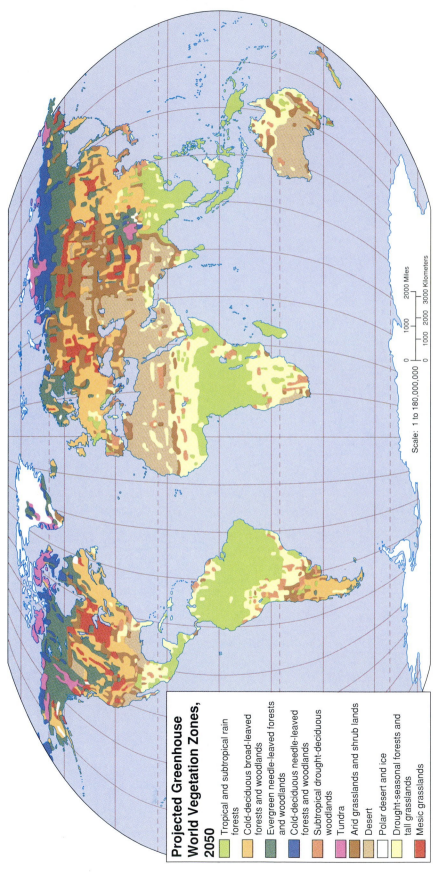

Projected Greenhouse World Vegetation Zones, 2050

- Tropical and subtropical rain forests
- Cold-deciduous broad-leaved forests and woodlands
- Evergreen needle-leaved forests and woodlands
- Cold-deciduous needle-leaved forests and woodlands
- Subtropical drought-deciduous woodlands
- Tundra
- Arid grasslands and shrub lands
- Desert
- Polar desert and ice
- Drought-seasonal forests and tall grasslands
- Mesic grasslands

Scale: 1 to 180,000,000

0   1000   2000 Miles

0   1000   2000   3000 Kilometers

A warmer world could result from the continued increase in atmospheric carbon dioxide and other gases that enhance the atmosphere's natural ability to retain the heat that is released from the earth. While most scenarios of global warming have to do with the specter of rising seas flooding coastal cities, or with scorching heat waves and droughts altering the patterns of agricultural production, the pattern of vegetation change shown here is at once more likely and more important. Patterns of human food production depend essentially upon the relationship between climate, soil, and vegetation. As indicated on this map, those patterns would change on an earth some 4 degrees (Celsius) warmer than at present. Three trends are most evident: an expansion of tropical forests, an increase in the areal

coverage of needleleaf forests, and a considerable growth in size in the world's drylands of desert and steppe. Growth in these three types of vegetative communities suggests shrinkage in some others. The problem for humans is that those vegetative communities that are likely to expand are those least suitable for human use as agricultural lands. The soils that develop under tropical forests are notoriously poor for farming, wearing out quickly under cultivation. The soils of coniferous forests are high in acids and are also only marginally useful for farming. And desert areas require costly irrigation technologies in order to be productive. As these three types of vegetation communities expand in a greenhouse world, the net result could be an earth that is less capable of producing food than at present.

Table I

Wood Production and Trade (in thousands of cubic meters)

	ROUNDWOOD PRODUCTION			PROCESSED WOOD PRODUCTION			
	TOTAL	FUEL AND CHARCOAL	INDUSTRIAL ROUNDWOOD	SAWNWOOD	PANELS	TOTAL PAPER PRODUCTION	ANNUAL NET TRADE IN ROUNDWOOD[a] (1991–1993)
AFRICA	**539,683**	**480,752**	**58,931**	**8,219**	**1,593**	**2,558**	**−4,216**
Algeria	2,305	2,006	298	13	50	91	X
Angola	6,382	5,483	899	5	11	0	X
Benin	5,374	5,075	299	25	X	X	X
Botswana	1,399	1,312	86	X	X	X	X
Burkina Faso	9,254	8,836	418	2	X	X	X
Burundi	4,484	4,431	53	3	X	X	X
Cameroon	14,483	11,490	2,993	481	78	5	−732
Central African Republic	3,628	3,185	443	63	2	X	X
Chad	4,164	3,569	595	2	X	X	X
Congo	3,438	2,155	1,283	53	36	X	X
Egypt	2,352	2,243	109	X	76	203	74
Equatorial Guinea	613	447	166	10	7	X	X
Ethiopia[b]	44,937	43,227	1,711	12	9	5	0
Gabon	4,345	2,712	1,633	32	140	X	X
Gambia	948	927	21	1	X	X	X
Ghana	16,965	15,512	1,453	407	60	X	X
Guinea	4,359	3,792	567	66	0	X	−15
Guinea-Bissau	572	422	150	16	X	X	X
Ivory Coast	13,370	10,498	2,872	606	229	X	−323
Kenya	37,324	35,513	1,811	185	52	148	0
Lesotho	635	635	X	X	X	X	X
Liberia	6,000	5,040	960	411	8	X	−234
Libya	646	536	110	31	X	6	−1
Madagascar	8,600	7,793	807	235	5	5	X
Malawi	9,730	9,235	496	44	17	X	X
Mali	5,955	5,575	380	13	X	X	X
Mauritania	13	8	5	X	X	X	X
Mauritius	15	2	13	5	0	X	16
Morocco	2,296	1,426	869	83	34	106	376
Mozambique	15,980	15,022	958	21	4	2	−7
Namibia	X	X	X	X	X	X	X
Niger	5,292	4,965	326	2	X	X	X
Nigeria	114,704	106,441	8,263	2,723	109	63	−49
Rwanda	5,647	5,392	255	27	2	X	X
Senegal	4,953	4,281	672	23	X	X	X
Sierra Leone	3,227	3,106	121	9	X	X	X
Somalia	8,761	8,655	106	14	0	X	0
South Africa	19,747	7,146	12,601	1,792	350	1,735	−1,066
Sudan	24,108	21,877	2,231	3	2	3	X
Swaziland	2,297	560	1,737	103	8	X	0
Tanzania	34,911	32,849	2,062	156	15	25	−9
Togo	1,265	1,072	192	3	X	X	−1
Tunisia	3,324	3,169	155	10	102	67	40
Uganda	15,099	13,103	1,996	64	4	3	0
Zaire	43,252	40,101	3,151	105	28	2	−111
Zambia	13,778	12,952	826	106	81	3	−4
Zimbabwe	8,033	6,269	1,764	250	79	86	−6
NORTH AND MIDDLE AMERICA	**730,441**	**155,803**	**574,637**	**165,733**	**38,264**	**94,943**	**−25,723**
Belize	188	126	62	14	X	X	−2
Canada	172,703	6,834	165,869	56,044	6,538	16,900	601
Costa Rica	4,168	3,134	1,034	661	65	19	2
Cuba	3,142	2,531	611	130	149	78	X
Dominican Republic	982	976	6	0	X	9	2
El Salvador	6,362	6,216	146	70	X	17	X
Guatemala	11,263	11,142	121	78	16	14	1

(Continued on next page)

	ROUNDWOOD PRODUCTION			PROCESSED WOOD PRODUCTION			
	TOTAL	**FUEL AND CHARCOAL**	**INDUSTRIAL ROUNDWOOD**	**SAWNWOOD**	**PANELS**	**TOTAL PAPER PRODUCTION**	**ANNUAL NET TRADE IN ROUNDWOOD[a] (1991–1993)**
Haiti	6,052	5,813	239	14	X	X	0
Honduras	6,298	5,672	626	362	13	X	X
Jamaica	539	385	154	27	0	4	18
Mexico	22,940	15,449	7,491	2,696	645	2,828	−210
Nicaragua	3,569	3,269	300	69	5	X	0
Panama	1,018	910	109	30	18	28	6
Trinidad and Tobago	65	22	43	46	X	X	3
United States	491,000	93,300	397,700	105,489	30,816	75,045	−26,191
SOUTH AMERICA	**362,400**	**243,585**	**118,815**	**26,411**	**4,287**	**8,344**	**−7,611**
Argentina	11,660	4,447	7,214	1,298	326	972	−1,002
Bolivia	1,530	1,377	153	208	55	0	−1
Brazil	268,879	191,166	77,713	18,628	2,552	5,051	−443
Chile	29,066	8,979	20,087	3,117	492	537	−6,005
Colombia	20,619	16,936	3,683	813	177	582	−1
Ecuador	7,435	4,218	3,218	892	220	150	X
Guyana	175	14	161	12	2	X	−9
Paraguay	8,502	5,396	3,106	313	127	13	X
Peru	8,031	6,981	1,050	511	32	327	1
Suriname	149	19	131	42	7	X	−2
Uruguay	4,015	3,034	981	248	7	80	−129
Venezuela	2,086	954	1,132	312	292	633	13
ASIA[c]	**1,122,978**	**849,658**	**273,320**	**99,785**	**39,094**	**63,606**	**48,013**
Afghanistan	7,327	5,683	1,644	400	1	X	−3
Armenia	X	X	X	X	X	X	X
Azerbaijan	X	X	X	X	X	X	X
Bangladesh	31,751	31,021	730	79	9	97	X
Bhutan	1,460	1,333	127	35	13	X	X
Cambodia	6,782	5,726	1,057	127	2	0	X
China	291,046	196,149	94,897	21,702	11,819	20,757	6,449
Georgia	X	X	X	X	X	X	X
India	282,384	257,813	24,571	17,460	442	2,505	571
Indonesia	185,426	146,342	39,084	8,471	10,359	2,206	−1,466
Iran	7,405	2,511	4,894	177	304	200	41
Iraq	153	103	50	8	3	13	X
Israel	113	13	100	X	177	209	X
Japan	34,110	372	33,738	27,267	8,228	28,380	46,485
Jordan	11	7	4	X	X	20	X
Kazakhstan	X	X	X	X	X	X	X
Korea, North	4,783	4,183	600	280	X	80	−89
Korea, South	6,485	4,491	1,994	3,584	1,634	5,410	9,327
Kuwait	X	X	X	X	X	X	58
Kyrgyzstan	X	X	X	X	X	X	X
Laos	4,681	4,132	549	231	13	X	X
Lebanon	487	479	8	10	46	42	30
Malaysia	52,906	9,157	43,750	9,249	3,288	531	−15,681
Mongolia	1,063	535	528	173	3	X	X
Myanmar	22,566	18,715	3,851	308	16	12	−1,352
Nepal	19,595	18,975	620	620	X	13	X
Oman	X	X	X	X	X	X	0
Pakistan	27,019	24,379	2,640	1,497	80	214	79
Philippines	39,137	35,149	3,988	631	433	484	256
Saudi Arabia	X	X	X	X	X	X	89
Singapore	160	160	X	27	361	85	−218
Sri Lanka	9,247	8,588	659	5	10	26	−16
Syria	74	23	52	9	27	1	−1
Tajikistan	X	X	X	X	X	X	X
Thailand	37,619	34,854	2,765	910	611	1,138	1,466
Turkey	15,317	9,750	5,567	5,133	1,121	924	1,539
Turkmenistan	X	X	X	X	X	X	X

(Continued on next page)

	ROUNDWOOD PRODUCTION			PROCESSED WOOD PRODUCTION			
	TOTAL	**FUEL AND CHARCOAL**	**INDUSTRIAL ROUNDWOOD**	**SAWNWOOD**	**PANELS**	**TOTAL PAPER PRODUCTION**	**ANNUAL NET TRADE IN ROUNDWOOD[a] (1991–1993)**
United Arab Emirates	X	X	X	X	X	X	36
Uzbekistan	X	X	X	X	X	X	X
Vietnam	33,008	28,407	4,601	861	39	116	−193
Yemen	X	X	X	X	X	X	X
EUROPE[c]	**319,100**	**50,672**	**268,428**	**76,383**	**34,710**	**68,233**	**13,669**
Albania	2,556	1,556	1,000	382	16	44	X
Austria	13,759	2,860	10,899	7,015	1,896	3,214	5,205
Belarus	10,714	819	9,895	1,619	525	221	X
Belgium	4,412	550	3,862	1,204	2,503	1,176	1,424
Bosnia-Herzegovina	X	X	X	X	X	X	−2
Bulgaria	3,599	1,695	1,904	564	268	188	−101
Croatia	2,131	710	1,421	671	90	X	−270
Czech Republic	10,306	996	9,310	2,650	739	656	−1,425
Denmark	2,245	493	1,752	688	294	330	−80
Estonia	2,293	926	1,365	300	140	42	X
Finland	37,663	3,320	34,343	7,111	1,044	9,304	5,144
France	43,617	10,450	33,167	10,181	3,514	7,652	−2,421
Germany	36,245	3,795	32,450	13,295	9,067	12,998	−5,825
Greece	2,726	1,502	1,224	354	372	508	102
Hungary	5,094	2,310	2,784	717	387	335	−945
Iceland	X	X	X	X	X	X	13
Ireland	1,795	50	1,745	465	244	36	−385
Italy	9,240	4,750	4,490	1,797	3,670	5,951	6,719
Latvia	3,515	700	2,815	593	190	28	X
Lithuania	X	X	X	X	200	50	−187
Macedonia	X	X	X	X	X	X	X
Moldova	X	X	X	X	X	X	X
Netherlands	1,397	156	1,241	400	105	2,851	40
Norway	10,516	934	9,582	2,329	564	1,812	749
Poland	18,314	2,975	15,339	3,822	1,666	1,128	−981
Portugal	11,409	598	10,811	1,176	1,085	904	−43
Romania	12,110	2,396	9,714	2,384	835	359	62
Russia	244,488	56,738	187,750	46,685	6,909	5,115	−11,098
Slovakia	5,003	521	4,482	80	X	X	−374
Slovenia	1,470	314	1,157	422	327	407	−106
Spain	15,216	1,990	13,226	2,784	2,361	3,458	1,800
Sweden	59,907	4,424	55,483	12,110	912	8,503	4,266
Switzerland	4,571	825	3,746	1,554	856	1,299	−189
Ukraine	X	X	X	X	X	X	2
United Kingdom	6,197	263	5,934	2,139	1,907	5,115	377
Yugoslavia	X	X	X	X	X	X	−299
OCEANIA	**44,140**	**8,748**	**35,392**	**5,827**	**1,766**	**2,803**	**−13,521**
Australia	19,860	2,896	16,964	3,028	901	2,007	−6,470
Fiji	307	37	270	108	17	X	−230
New Zealand	15,110	50	15,060	2,527	802	796	−4,685
Papua New Guinea	8,188	5,533	2,655	117	46	X	X
Solomon Islands	468	138	330	16	0	X	X

[a]Imports of roundwood are shown as positive numbers; exports are shown as negative numbers.

[b]Totals for Ethiopia include Eritrea.

[c]Regional totals for Asia and Europe do not include former republics of the USSR.

Source: World Resources 1994–95.

Table J

Globally Threatened Animal Species, 1990

	MAMMALS		BIRDS		REPTILES		AMPHIBIANS		FRESHWATER FISH
	THREATENED SPECIES	THREATENED SPECIES PER 10,000 KM²	THREATENED SPECIES	THREATENED SPECIES PER 10,000 KM²	THREATENED SPECIES	THREATENED SPECIES PER 10,000 KM²	THREATENED SPECIES	THREATENED SPECIES PER 10,000 KM²	THREATENED SPECIES PER 10,000 KM²
AFRICA									
Algeria	12	2	15	2	0	0¹	0	0¹	1
Angola	14	3	12	2	2	0	0	0	0¹
Benin	11	5	1	0¹	2	1	0	0	0
Botswana	9	2	6	2	1	0	0	0	0
Burkina Faso	10	3	1	0	2	1	0	0	0
Burundi	4	3	5	4	1	1	0	0	0
Cameroon	27	8	17	5	2	1	1	0	11
Central African Republic	12	3	2	1	2	1	0	0	0
Chad	18	4	4	1	2	0	0	0	0
Congo	12	4	3	1	2	1	0	0	0
Egypt	9	2	16	3	2	0	0	0	1
Equatorial Guinea	15	11	3	2	2	1	1	1	0
Eritrea	X	X	X	X	X	X	X	X	X
Ethiopia	25	5	14	3	1	0	0	0	0
Gabon	14	5	4	1	2	1	0	0	0
Gambia	7	7	1	1	2	2	0	0	0
Ghana	13	5	8	3	2	1	0	0	0
Guinea	17	6	6	2	1	0	1	0	0
Guinea-Bissau	5	3	2	1	2	1	0	0	0
Ivory Coast	18	6	9	3	1	0	1	0	0
Kenya	15	4	18	5	2	1	1	0	0
Lesotho	2	1	7	5	0	0	0	0	0
Liberia	18	8	10	5	2	1	0	0	0
Libya	12	2	9	2	1	0	0	0	0
Madagascar	53	14	28	7	10	3	0	0	0
Malawi	10	4	7	3	1	0	0	0	0
Mali	16	3	4	1	2	0	0	0	0
Mauritania	14	3	5	1	1	0	0	0	0
Mauritius	3	5	10	17	6	10	0	0	1
Morocco	9	3	14	4	0	0	0	0	1
Mozambique	10	2	11	3	1	0	0	0	1
Namibia	11	3	7	2	2	0	0	0	4
Niger	15	3	1	0	1	0	0	0	0
Nigeria	25	6	10	2	2	1	0	0	0
Rwanda	11	8	7	5	2	1	0	0	0
Senegal	11	4	5	2	2	1	0	0	0
Sierra Leone	13	7	7	4	2	1	0	0	0
Somalia	16	4	7	2	1	0	0	0	0
South Africa	26	5	13	3	3	1	1	0	28
Sudan	17	3	8	1	1	0	0	0	0
Swaziland	0	0¹	5	4	1	1	0	0	0
Tanzania	30	7	26	6	3	1	0	0	0
Togo	9	5	1	1	2	1	0	0	0

(Continued on next page)

	MAMMALS		BIRDS		REPTILES		AMPHIBIANS		FRESHWATER FISH	
	THREATENED SPECIES	THREATENED SPECIES PER 10,000 KM²	THREATENED SPECIES	THREATENED SPECIES PER 10,000 KM²	THREATENED SPECIES	THREATENED SPECIES PER 10,000 KM²	THREATENED SPECIES	THREATENED SPECIES PER 10,000 KM²	THREATENED SPECIES	THREATENED SPECIES PER 10,000 KM²
Tunisia	6	2	14	6	1	0	0	0	0	0
Uganda	16	6	12	4	1	0	0	0	0	0
Zaire	22	4	27	4	2	0	0	0	1	0
Zambia	10	2	10	2	2	0	0	0	0	0
Zimbabwe	9	3	6	2	1	0	0	0	0	0
NORTH AND MIDDLE AMERICA										
Belize	8	6	4	3	3	2	0	0	0	0
Canada	5	1	6	1	0	0	0	0	15	0
Costa Rica	10	6	14	8	2	1	0	0	0	0
Cuba	2	1	15	7	4	2	0	0	0	0
Dominican Republic	1	1	5	3	4	2	0	0	0	0
El Salvador	6	5	2	2	1	1	0	0	0	0
Guatemala	10	5	10	5	4	2	0	0	0	0
Haiti	1	1	4	3	4	3	0	0	0	0
Honduras	7	3	11	5	3	1	0	0	0	0
Jamaica	6	6	2	2	3	3	0	0	0	0
Mexico	26	5	35	6	16	3	4	1	98	0
Nicaragua	8	3	7	3	2	1	0	0	0	0
Panama	13	7	14	7	2	1	0	0	0	0
Trinidad and Tobago	1	1	3	4	0	0	0	0	0	0
United States	21	2	43	4	25	3	22	2	164	2
SOUTH AMERICA										
Argentina	25	4	53	8	4	1	1	0	1	0
Bolivia	21	4	34	7	4	1	0	0	1	0
Brazil	24	3	123	13	11	1	0	0	9	0
Chile	9	2	18	4	0	0	0	0	1	0
Colombia	25	5	69	14	10	2	0	0	0	0
Ecuador	21	7	64	21	8	3	0	0	0	0
Guyana	12	4	9	3	3	1	0	0	1	0
Paraguay	14	4	24	7	4	1	0	0	0	0
Peru	29	6	65	13	6	1	1	0	1	0
Suriname	11	4	6	2	1	0	0	0	0	0
Uruguay	5	2	11	4	2	1	0	0	0	0
Venezuela	19	4	34	8	2	0	0	0	0	0
ASIA										
Afghanistan	13	3	13	3	1	1	1	1	0	0
Armenia	X	X	X	X	X	X	X	X	X	X
Azerbaijan	X	X	X	X	X	X	X	X	X	X
Bangladesh	15	6	27	11	14	6	X	X	X	X
Bhutan	15	9	10	6	1	1	0	0	0	0
Cambodia	21	8	13	5	6	2	0	0	5	0
China	30	3	83	9	7	1	1	1	7	0
Georgia	X	X	X	X	X	X	X	X	X	X
India	38	6	72	11	17	3	3	3	2	0

(Continued on next page)

(Continued on next page)

	MAMMALS		BIRDS		REPTILES		AMPHIBIANS		FRESHWATER FISH
	THREATENED SPECIES	THREATENED SPECIES PER 10,000 KM²	THREATENED SPECIES	THREATENED SPECIES PER 10,000 KM²	THREATENED SPECIES	THREATENED SPECIES PER 10,000 KM²	THREATENED SPECIES	THREATENED SPECIES PER 10,000 KM²	THREATENED SPECIES PER 10,000 KM²
Indonesia	50	9	135	24	13	2	0	0	29
Iran	15	3	20	4	4	1	0	0	2
Iraq	9	3	17	5	0	0	1	0	2
Israel	8	6	15	12	1	1	1	1	0
Japan	5	2	31	9	0	0	0	0	3
Jordan	5	2	11	5	0	0	1	1	0
Kazakhstan	X	X	X	X	X	X	X	X	X
Korea, North	5	2	25	11	0	0	0	0	0
Korea, South	6	3	22	10	0	0	0	0	0
Kuwait	5	4	7	6	0	0	0	0	0
Kyrgyzstan	X	X	X	X	X	X	X	X	X
Laos	23	8	18	6	5	2	0	0	5
Lebanon	4	4	15	15	1	1	0	0	0
Malaysia	23	7	35	11	12	4	0	0	6
Mongolia	9	2	13	2	0	0	0	0	0
Myanmar	23	6	42	10	10	2	0	0	2
Nepal	22	9	20	8	9	4	0	0	0
Oman	5	2	8	3	9	0	0	0	2
Pakistan	15	4	25	6	6	1	0	0	0
Philippines	12	4	39	13	6	2	0	0	21
Saudi Arabia	9	2	12	2	0	0	0	0	0
Singapore	X	X	X	X	X	X	X	X	X
Sri Lanka	7	4	8	4	3	2	0	0	12
Syria	4	2	15	6	1	0	0	0	0
Tajikistan	X	X	X	X	X	X	X	X	X
Thailand	26	7	34	9	9	2	0	0	13
Turkey	5	1	18	4	5	1	1	0	5
Turkmenistan	X	X	X	X	X	X	X	X	X
United Arab Emirates	4	2	7	3	0	0	0	0	0
Uzbekistan	X	X	X	X	X	X	X	X	X
Vietnam	28	9	34	11	8	3	1	0	4
Yemen	X	X	X	X	X	X	X	X	X
EUROPE									
Albania	2	1	14	10	1	1	0	0	1
Austria	2	1	13	6	0	0	0	0	2
Belarus	X	X	X	X	X	X	X	X	X
Belgium	2	1	13	9	0	0	0	0	1
Bulgaria	3	1	15	7	1	0	0	0	3
Czechoslovakia (former)	2	1	18	8	0	0	0	0	2
Denmark	1	1	16	10	0	0	0	0	0
Estonia	X	X	X	X	X	X	X	X	X
Finland	3	1	12	4	0	0	0	0	1
France	6	2	21	6	2	1	0	0	3
Germany	2	1	17	5	0	0	0	0	3
Greece	4	2	19	8	3	1	0	0	6
Hungary	2	1	16	8	0	0	0	0	2

	MAMMALS		BIRDS		REPTILES		AMPHIBIANS		FRESHWATER FISH
	THREATENED SPECIES	THREATENED SPECIES PER 10,000 KM²	THREATENED SPECIES	THREATENED SPECIES PER 10,000 KM²	THREATENED SPECIES	THREATENED SPECIES PER 10,000 KM²	THREATENED SPECIES	THREATENED SPECIES PER 10,000 KM²	THREATENED SPECIES PER 10,000 KM²
Iceland	1	0	2	1	0	0	0	0	1
Ireland	0	0	10	5	0	0	0	0	1
Italy	3	1	19	6	2	1	7	2	3
Latvia	X	X	X	X	X	X	X	X	X
Lithuania	X	X	X	X	X	X	X	X	X
Moldova	X	X	X	X	X	X	X	X	X
Netherlands	2	1	13	8	0	0	0	0	1
Norway	3	1	8	3	0	0	0	0	1
Poland	4	1	16	5	0	0	0	0	1
Portugal	6	3	18	9	0	0	1	0	0
Romania	2	1	18	6	1	0	0	0	4
Russia	20	2	38	3	3	0	0	0	5
Slovakia	X	X	X	X	X	X	X	X	X
Spain	6	2	23	6	5	1	3	1	2
Sweden	1	0	14	4	0	0	0	0	1
Switzerland	2	1	15	9	0	0	1	1	3
Ukraine	X	X	X	X	X	X	X	X	X
United Kingdom	2	1	22	8	0	0	0	0	1
Yugoslavia (former)	3	1	17	6	1	0	2	1	5
OCEANIA									
Australia	35	4	39	4	9	1	3	0	16
Fiji	1	1	5	4	4	3	1	1	0
New Zealand	1	0	26	9	1	0	3	1	2
Papua New Guinea	4	1	25	7	1	0	0	0	0
Solomon Islands	2	1	20	14	3	2	0	0	0

[1]In this and other columns showing threatened species per 10,000 km², 0 indicates a figure of less than 0.5 threatened species per 10,000 km². In columns showing the total number of threatened species, 0 is an absolute figure.

Source: World Resources 1992–93.

Table K
Rare and Threatened Plants, 1991

	NUMBER OF PLANT TAXA	NUMBER OF RARE AND THREATENED PLANT TAXA	RARE AND THREATENED PLANT TAXA PER 1,000 EXISTING TAXA	RARE AND THREATENED PLANT TAXA PER 10,000 KM^2
AFRICA				
Algeria	3,139–3,150	144	46	24
Angola	5,000	19	4	4
Benin	2,000	3	2	1
Botswana	2,600–2,800	4	1–2	1
Burkina Faso	1,096	0	0[1]	0[2]
Burundi	2,500	0	0	0
Cameroon	8,000	74	9	21
Central African Republic	3,600	0	0	0
Chad	1,600	14	9	3
Congo	4,000	4	1	1
Egypt	2,085	93	45	20
Equatorial Guinea	X	X	X	X
Eritrea	X	X	X	X
Ethiopia	6,283	44	7	9
Gabon	8,000	80	10	27
Gambia	530	0	0	0
Ghana	3,600	34	9	12
Guinea	X	36	X	13
Guinea-Bissau	1,000	0	0	0
Ivory Coast	3,660	70	19	22
Kenya	6,500	144	22	38
Lesotho	1,591	7	4	5
Liberia	X	1	X	0
Libya	1,600–1,800	58	32–36	11
Madagascar	10,000–12,000	193	16–19	50
Malawi	3,600	61	17	29
Mali	1,600	15	9	3
Mauritania	1,100	3	3	1
Mauritius	800–900	240	267–300	419
Morocco	3,500–3,600	194	54–55	55
Mozambique	5,500	84	15	20
Namibia	3,159	18	6	4
Niger	1,178	1	1	0
Nigeria	4,614	9	2	2
Rwanda	2,150	0	0	0
Senegal	2,100	32	15	12
Sierra Leone	2,480	12	5	6
Somalia	3,000	51	17	13
South Africa	23,000	1,145	50	235
Sudan	3,200	9	3	1
Swaziland	2,715	25	9	21
Tanzania	10,000	158	16	36
Togo	2,302	0	0	0
Tunisia	2,120–2,200	26	12	11
Uganda	5,000	11	2	4
Zaire	11,000	3	0	1
Zambia	4,600	1	0	0
Zimbabwe	5,428	96	18	29
NORTH AND MIDDLE AMERICA				
Belize	3,240	38	12	29
Canada	3,220	13	4	1
Costa Rica	8,000	456	57	266
Cuba	7,000	874	125	396
El Salvador	2,500	24	10	19
Guatemala	8,000	305	38	139
Hispaniola (Haiti and Dominican Republic)	5,000	0	0	0
Honduras	5,000	48	10	22
Jamaica	3,582	8	2	8
Mexico	20,000	1,111	56	196

(Continued on next page)

	NUMBER OF PLANT TAXA	NUMBER OF RARE AND THREATENED PLANT TAXA	RARE AND THREATENED PLANT TAXA PER 1,000 EXISTING TAXA	RARE AND THREATENED PLANT TAXA PER 10,000 KM^2
Nicaragua	5,000	72	14	32
Panama	8,000–9,000	344	38–43	176
Trinidad and Tobago	2,281	4	2	5
United States	20,000	2,476	124	261
SOUTH AMERICA				
Argentina	9,000	157	17	25
Bolivia	15,000–18,000	31	2	7
Brazil	55,000	240	4	26
Chile	5,500	192	35	46
Colombia	45,000	316	7	68
Ecuador	10,000–20,000	121	6–12	40
Guyana	6,000–8,000	68	9–11	25
Paraguay	7,000–8,000	12	2	4
Peru	20,000	353	18	71
Suriname	4,500	68	15	27
Uruguay	X	11	X	4
Venezuela	15,000–20,000	105	5–7	24
ASIA				
Afghanistan	3,000	2	1	1
Armenia	X	X	X	X
Azerbaijan	X	X	X	X
Bangladesh	5,000	6	1	3
Bhutan	5,000	6	1	4
Cambodia	X	11	X	4
China	30,000	841	28	88
Georgia	X	X	X	X
India	15,000	1,349	90	206
Indonesia	X	X	X	X
Iran	7,000	1	0	0
Iraq	2,937	3	1	1
Israel	2,317	39	17	31
Japan	4,022	687	171	207
Jordan	2,200	13	6	6
Kazakhstan	X	X	X	X
Korea, North	X	X	X	X
Korea, South	2,838	0	0	0
Kuwait	350	1	3	1
Kyrgyzstan	X	X	X	X
Laos	X	3	X	1
Lebanon	3,000	6	2	6
Malaysia	X	X	X	X
Mongolia	X	X	X	X
Myanmar	7,000	23	3	6
Nepal	6,500	21	3	9
Oman	1,100	2	2	1
Pakistan	5,500–6,000	8	1	2
Philippines	8,900	106	12	35
Saudi Arabia	3,500	1	0	0
Singapore	2,030	16	8	40
Sri Lanka	3,700	209	56	113
Syria	3,000	13	4	5
Tajikistan	X	X	X	X
Thailand	12,000	63	5	17
Turkey	10,150	1,952	192	466
Turkmenistan	X	X	X	X
United Arab Emirates	X	X	X	X
Uzbekistan	X	X	X	X
Vietnam	8,000	388	49	123
Yemen	1,000	1	1	0
EUROPE				
Albania	3,100–3,300	76	23–25	54
Austria	2,900–3,100	25	8–9	12

(Continued on next page)

	NUMBER OF PLANT TAXA	NUMBER OF RARE AND THREATENED PLANT TAXA	RARE AND THREATENED PLANT TAXA PER 1,000 EXISTING TAXA	RARE AND THREATENED PLANT TAXA PER 10,000 KM^2
Belarus	X	X	X	X
Belgium	1,600–1,800	11	6–7	8
Bulgaria	3,500–3,650	89	24–25	40
Czechoslovakia (former)	2,600–2,750	29	11	13
Denmark	1,000	7	7	4
Estonia	X	X	X	X
Finland	1,150–1,450	7	5–6	2
France	4,300–4,450	112	25–26	30
Germany	2,476	16	6	6
Greece	5,000	531	106	227
Hungary	2,400	21	9	10
Iceland	470	2	4	1
Ireland	1,000–1,150	4	4	2
Italy	4,750–4,900	151	31–32	49
Latvia	X	X	X	X
Lithuania	X	X	X	X
Moldova	X	X	X	X
Netherlands	1,400	7	5	5
Norway	1,600–1,800	12	7–8	4
Poland	2,250–2,450	16	7	5
Portugal	2,400–2,600	90	35–38	43
Romania	3,300–3,400	68	20–21	24
Russia	21,000	531	25	42
Spain	4,750–4,900	449	92–95	124
Sweden	1,600–1,800	9	5–6	3
Switzerland	2,600–2,750	19	7	12
Ukraine	X	X	X	X
United Kingdom	1,700–1,850	22	12–13	8
Yugoslavia (former)	4,750–4,900	191	39–40	66
OCEANIA				
Australia	18,000	2,133	119	239
Fiji	1,500	25	17	20
New Zealand	2,000	254	127	86
Papua New Guinea	X	68	X	19
Solomon Islands	2,150	3	1	2

[1]Less than 0.5 rare and threatened plant taxa per 1,000 existing taxa.

[2]Less than 0.5 rare and threatened plant taxa per 10,000 km^2.

Source: World Resources 1992–93.

Table L

Habitat Extent and Loss (in thousands of hectares), 1980–1990

| | FORESTS | | | | | | SAVANNA/ GRASSLAND | | DESERT/SCRUB | | WETLANDS/ MARSH | | MANGROVES | |
| | ALL FORESTS | | DRY FORESTS | | MOIST FORESTS | | | | | | | | | |
	CURRENT EXTENT	% LOST	CURRENT EXTENT	% LOST	CURRENT EXTENT	% LOST	CURRENT EXTENT	% LOST	CURRENT EXTENT	% LOST	CURRENT EXTENT	% LOST	CURRENT EXTENT	% LOST
AFRICA	X	X	X	X	X	X	X	X	X	X	X	X	X	X
Algeria	X	X	X	X	X	X	X	X	X	X	730	X	0	0
Angola	51,428	45	40,261	45	11,167	48	24,590	17	456	20	X	X	110	50
Benin	4,448	62	4,406	55	42	97	0	0	0	0	0	0	7	X
Botswana	11,293	62	11,293	62	0	0	12,247	53	0	0	2,331	10	0	0
Burkina Faso	4,964	80	4,964	80	0	0	768	70	0	0	0	X	0	0
Burundi	117	91	114	91	3	95	246	80	0	0	14	80	0	0
Cameroon	18,468	59	2,949	69	15,519	56	376	72	0	0	16	0	486	40
Central African Republic	27,933	55	14,667	51	13,266	59	0	0	0	0	0	0	0	0
Chad	5,848	80	5,848	80	0	0	11,958	72	0	0	66	90	0	0
Congo	17,442	49	0	0	17,442	49	0	0	0	0	290	X	2	0
Egypt	X	X	0	X	0	X	0	X	X	X	809	0	X	X
Equatorial Guinea	1,285	50	0	0	1,285	50	0	0	0	0	0	X	12	60
Eritrea	X	X	X	X	X	X	X	X	X	X	X	X	X	X
Ethiopia	5,570	86	5,570	86	0	0	27,469	61	525	30	0	0	0	0
Gabon	17,245	35	0	0	17,245	35	0	0	0	0	0	0	115	50
Gambia	122	89	72	90	50	88	0	0	0	0	0	X	50	X
Ghana	4,254	82	2,670	71	1,584	89	0	0	0	0	853	X	2	X
Guinea	7,440	69	1,799	71	5,641	69	0	0	0	0	525	0	120	60
Guinea-Bissau	512	80	0	0	512	80	0	0	0	0	0	X	315	70
Ivory Coast	6,308	80	3,562	60	2,746	88	27,682	43	0	0	32	0	3	X
Kenya	2,274	71	2,130	67	144	90	141	70	0	0	0	0	93	70
Lesotho	851	67	851	67	0	0	0	0	0	0	0	X	0	0
Liberia	1,424	87	8	20	1,416	87	0	0	0	0	40	0	36	70
Libya	X	X	X	X	X	X	X	X	X	X	X	X	0	X
Madagascar	13,049	75	11,401	62	1,648	84	1,509	78	0	0	197	X	130	40
Malawi	3,977	56	3,977	56	0	0	0	0	0	0	112	60	0	0
Mali	7,670	78	7,670	78	0	0	8,368	80	0	0	2,000	X	0	0
Mauritania	6	90	6	90	0	0	4,610	88	0	0	0	0	0	0
Mauritius	11	X	X	X	X	X	X	X	X	X	0	0	X	X
Morocco	X	X	X	X	X	X	X	X	X	X	33	X	0	0
Mozambique	33,137	57	33,137	57	0	0	696	20	0	0	171	10	276	60
Namibia	15,020	52	15,020	52	0	0	14,741	59	14,570	0	225	10	0	0
Niger	2,278	80	2,278	80	0	0	10,985	75	0	0	38	80	0	0
Nigeria	18,201	80	14,339	70	3,862	91	498	80	0	0	42	80	1,052	X
Rwanda	184	80	184	80	0	0	157	90	0	0	80	X	0	X
Senegal	2,455	82	2,250	80	205	93	1,120	80	0	0	2	X	185	X
Sierra Leone	554	92	48	40	506	93	0	0	0	0	0	0	102	X
Somalia	642	67	642	67	0	0	36,374	40	712	4	0	0	54	70
South Africa	20,444	46	20,444	46	0	0	32,257	62	880	0	0	0	45	50
Sudan	15,367	74	15,162	73	205	91	36,007	68	0	0	11,170	X	0	0
Swaziland	772	56	772	56	0	0	0	0	0	0	0	0	0	0

(Continued on next page)

	ALL FORESTS		DRY FORESTS		MOIST FORESTS		SAVANNA/GRASSLAND		DESERT/SCRUB		WETLANDS/MARSH		MANGROVES	
	CURRENT EXTENT	% LOST	CURRENT EXTENT	% LOST	CURRENT EXTENT	% LOST	CURRENT EXTENT	% LOST	CURRENT EXTENT	% LOST	CURRENT EXTENT	% LOST	CURRENT EXTENT	% LOST
Tanzania	36,137	40	35,867	39	270	80	14,352	49	0	0	1,545	X	212	60
Togo	1,758	69	1,622	57	136	92	0	0	0	0	0	0	0	X
Tunisia	X	X	X	X	X	X	X	X	X	X	868	X	0	0
Uganda	3,371	79	2,062	67	1,309	86	1,042	71	0	0	1,420	X	0	0
Zaire	83,255	57	9,135	54	74,120	57	5,405	30	0	0	215	50	125	50
Zambia	44,606	30	44,606	30	0	0	8,175	18	0	0	1,106	10	0	0
Zimbabwe	17,169	56	17,169	56	0	0	00	0	0	0	0	0	0	0
NORTH AND MIDDLE AMERICA	**X**	**X**	**X**	**X**	**X**	**X**	**X**	**X**	**X**	**X**	**X**	**X**	**X**	**X**
Belize	X	X	X	X	975	X	X	X	X	X	X	X	78	X
Canada	274,000	48	X	X	X	X	27,663	X	X	X	127,000	X	0	0
Costa Rica	X	X	X	X	1,540	X	X	X	X	X	82	X	35	0
Cuba	X	X	X	X	X	X	X	X	X	X	1,747	X	626	X
Dominican Republic	335	93	X	X	X	X	X	X	X	X	4,844	X	24	X
El Salvador	X	X	X	X	0	X	X	X	X	X	77	X	45	X
Guatemala	4,500	60	X	X	X	X	X	X	X	X	220	X	16	X
Haiti	X	X	X	X	X	X	X	X	X	X	113	X	18	X
Honduras	77	X	X	X	1,930	X	X	X	X	X	649	X	117	X
Jamaica	X	X	X	X	X	X	X	X	X	X	14	X	20	X
Mexico	38,461	66	X	X	X	X	X	X	100,000	X	3,264	X	1,420	X
Nicaragua	X	X	X	X	2,700	X	X	X	X	X	2,053	X	60	X
Panama	X	X	X	X	2,150	X	X	X	X	X	647	X	298	X
Trinidad and Tobago	X	X	X	X	X	X	X	X	X	X	21	X	9	X
United States	13,000	95	X	X	X	X	3,000	99	X	X	42,240	53	281	X
SOUTH AMERICA	**964,050**	**20**	**310,980**	**18**	**653,070**	**22**	**264,200**	**23**	**141,230**	**X**	**74,617**	**X**	**X**	**X**
Argentina	74,220	2	64,540	2	9,680	1	75,540	24	93,190	X	6,169	X	X	X
Bolivia	75,430	14	34,510	23	40,920	6	8,770	50	2,130	X	2,419	X	0	0
Brazil	524,190	28	155,590	17	368,600	31	74,000	20	0	X	29,690	X	X	X
Chile	20,930	22	7,520	28	13,410	18	10,110	12	27,370	X	8,827	X	X	X
Colombia	73,880	3	11,630	11	62,250	2	25,550	20	0	X	1,928	X	501	X
Ecuador	15,470	4	3,370	11	12,100	1	4,190	24	570	X	993	X	182	X
Guyana	17,700	2	540	5	17,160	1	1,840	8	0	X	814	X	80	X
Paraguay	21,800	19	20,910	20	890	2	10,400	20	0	X	5,724	X	0	0
Peru	74,270	12	8,800	47	65,470	3	13,900	41	15,230	X	1,303	X	6	X
Suriname	12,900	7	50	38	12,850	7	120	25	0	X	1,625	X	115	X
Uruguay	300	0	90	0	210	0	15,410	7	0	X	625	X	X	X
Venezuela	44,940	10	3,390	54	41,550	2	24,330	25	2,720	X	14,501	X	674	X
ASIA	**X**	**X**	**X**	**X**	**X**	**X**	**X**	**X**	**X**	**X**	**X**	**X**	**X**	**X**
Afghanistan	X	X	X	X	X	X	X	X	X	X	40	X	0	0
Armenia	X	X	X	X	X	X	X	X	X	X	X	X	0	X
Azerbaijan	X	X	X	X	X	X	0	X	0	X	X	X	X	X
Bangladesh	482	96	0	0	482	96	0	0	0	0	68	96	291	73

(Continued on next page)

	ALL FORESTS		DRY FORESTS		MOIST FORESTS		SAVANNA/GRASSLAND		DESERT/SCRUB		WETLANDS/MARSH		MANGROVES	
	CURRENT EXTENT	% LOST	CURRENT EXTENT	% LOST	CURRENT EXTENT	% LOST	CURRENT EXTENT	% LOST	CURRENT EXTENT	% LOST	CURRENT EXTENT	% LOST	CURRENT EXTENT	% LOST
Bhutan	2,298	33	700	30	1,598	35	0	0	0	0	7	X	0	0
Cambodia	3,885	78	1,608	81	2,277	74	0	0	0	0	389	45	16	5
China	6,000	99	X	X	X	X	X	X	391,680	X	4,200	X	67	X
Georgia	X	X	X	X	X	X	X	X	X	X	X	X	X	X
India	49,929	78	35,785	81	14,144	56	0	0	8,527	88	941	79	189	85
Indonesia	60,403	51	10,503	27	49,900	54	0	0	0	0	11,872	39	2,101	45
Iran	X	X	X	X	X	X	X	X	X	X	1,418	X	X	X
Iraq	X	X	X	X	X	X	X	X	X	X	1,921	X	0	X
Israel	X	X	X	X	X	X	X	X	X	X	170	X	0	0
Japan	1,204	X	X	X	X	X	X	X	X	X	250	X	0	0
Jordan	X	X	X	X	X	X	X	X	X	X	1	X	0	0
Kazakhstan	X	X	X	X	X	X	X	X	X	X	X	X	X	X
Korea, North	X	X	X	X	X	X	X	X	0	0	84	0	X	X
Korea, South	X	X	X	X	X	X	0	0	0	0	136	0	X	X
Kuwait	X	X	X	X	X	X	X	X	X	X	X	X	X	X
Kyrgyz	X	X	X	X	X	X	X	X	X	X	X	X	X	X
Laos	6,897	68	3,794	67	3,103	75	0	0	0	0	0	0	0	0
Lebanon	X	X	X	X	X	X	X	X	X	X	X	85	0	0
Malaysia	18,008	42	2,852	19	15,155	45	0	0	0	0	2,214	35	731	32
Mongolia	X	X	X	X	X	X	X	X	X	X	1,708	X	0	0
Myanmar	24,131	64	12,000	68	12,130	65	1,203	74	29	93	49	98	171	58
Nepal	5,381	54	882	16	4,499	58	0	0	0	0	291	X	0	X
Oman	X	X	X	X	X	X	X	X	X	X	X	X	X	X
Pakistan	764	86	184	96	580	27	0	0	2,811	69	320	74	154	78
Philippines	1,000[1]	X	X	X	X	X	X	X	X	X	1,322	X	140	X
Saudi Arabia	0	100	0	0	0	X	120,000	X	70,500	X	X	X	2	76
Singapore	0	100	0	0	0	100	0	0	0	0	0	0	0	X
Sri Lanka	610	86	446	76	163	94	X	X	495	75	512	0	120	X
Syria	X	X	X	X	X	X	X	X	X	X	38	X	0	X
Tajikistan	X	X	X	X	X	X	X	X	X	X	X	X	X	X
Thailand	13,107	73	8,330	78	4,777	57	X	X	0	0	83	96	19	87
Turkey	606	X	X	X	X	X	X	X	X	X	1,391	X	0	X
Turkmenistan	X	X	X	X	X	X	X	X	X	X	X	X	X	X
United Arab Emirates	X	X	X	X	X	X	X	X	X	X	X	X	3	X
Uzbekistan	X	X	X	X	X	X	X	X	X	X	X	X	X	X
Vietnam	6,758	76	2,105	68	4,654	79	0	0	0	0	26	100	147	62
Yemen	X	X	X	X	X	X	X	X	X	X	X	X	X	X
EUROPE	**X**	**X**	**X**	**X**	**X**	**X**	**X**	**X**	**X**	**X**	**X**	**X**	**X**	**X**
Albania	X	X	X	X	X	X	X	X	X	X	33	X	0	0
Austria	194	X	X	X	X	X	X	X	X	X	29	X	0	0
Belarus	X	X	X	X	X	X	X	X	X	X	X	X	0	0
Belgium	6	X	X	X	X	X	1	X	X	X	7	X	0	0
Bulgaria	165	X	X	X	X	X	X	X	X	X	15	X	0	0
Czechoslovakia (former)	135	X	X	X	X	X	X	X	X	X	69	X	0	0
Denmark	17	X	X	X	X	X	X	X	X	X	716	X	0	0

(Continued on next page)

| | FORESTS | | | | | | SAVANNA/ GRASSLAND | | DESERT/SCRUB | | WETLANDS/ MARSH | | MANGROVES | |
| | ALL FORESTS | | DRY FORESTS | | MOIST FORESTS | | | | | | | | | |
	CURRENT EXTENT	% LOST	CURRENT EXTENT	% LOST	CURRENT EXTENT	% LOST	CURRENT EXTENT	% LOST	CURRENT EXTENT	% LOST	CURRENT EXTENT	% LOST	CURRENT EXTENT	% LOST
Estonia	X	X	X	X	X	X	X	X	X	X	X	X	X	X
Finland	15	X	X	X	X	X	X	X	X	X	300	X	0	0
France	131	99	X	X	X	X	250	X	X	X	1,171	X	0	0
Germany	30	X	X	X	X	X	100	X	X	X	1,466	X	0	0
Greece	64	X	X	X	X	X	X	X	X	X	87	X	0	0
Hungary	267	X	X	X	X	X	200	X	X	X	94	X	0	0
Iceland	X	X	X	X	X	X	X	X	X	X	443	X	0	0
Ireland	3	X	X	X	X	X	700	X	X	X	115	X	0	0
Italy	1,215	X	X	X	X	X	200²	X	X	X	3,000	94	0	0
Latvia	X	X	X	X	X	X	X	X	X	X	X	X	X	X
Lithuania	X	X	X	X	X	X	X	X	X	X	X	X	X	X
Moldova	X	X	X	X	X	X	X	X	X	X	X	X	X	X
Netherlands	38	X	X	X	X	X	10	X	X	X	353	X	0	0
Norway	96	X	X	X	X	X	X	X	X	X	152	X	0	0
Poland	144	X	X	X	X	X	X	X	X	X	194	X	0	0
Portugal	X	X	X	X	X	X	755	X	X	X	85	X	0	0
Romania	193	X	X	X	X	X	X	X	X	X	483	X	0	0
Russia	37,573	38	X	X	X	X	X	X	X	X	2,837	X	X	X
Spain	582	X	X	X	X	X	1,452	X	X	X	445	X	0	0
Sweden	985	X	X	X	X	X	X	X	X	X	2,098	X	0	0
Switzerland	95	X	X	X	X	X	X	X	X	X	178	X	0	0
Ukraine	X	X	X	X	X	X	X	X	X	X	X	X	X	X
United Kingdom	200	X	X	X	X	X	5,298	X	X	X	446	X	0	0
Yugoslavia (former)	945	X	X	X	X	X	X	X	X	X	89	X	0	0
OCEANIA	**X**	**X**	**X**	**X**	**X**	**X**	**X**	**X**	**X**	**X**	**X**	**X**	**X**	
Australia	13,000	95	X	X	X	X	75,900	X	87,600	58	17,000	95	2,200	X
Fiji	750	X	X	X	X	X	X	X	X	X	X	X	42	7
New Zealand	5,000	77	18,807	69	X	X	6,500	90	X	X	32,240	90	20	X
Papua New Guinea	X	X	350	X	23,600	X	2,800	X	X	X	5,000	X	200	X
Solomon Islands	X	X	X	X	X	X	X	X	X	X	X	X	64	X

[1]Figure shown represents maximum current extent; actual extent is estimated to be lower than this figure.

[2]Figure shown represents minimum current extent.

Source: World Resources 1994–95.

Table M

National Protection of Natural Areas

	ALL PROTECTED AREAS			TOTALLY PROTECTED AREAS		PARTIALLY PROTECTED AREAS		PERCENT OF PROTECTED AREAS	
	NUMBER	AREA (thousands of hectares)	% OF LAND AREA	NUMBER	AREA (thousands of hectares)	NUMBER	AREA (thousands of hectares)	AT LEAST 100,000 HECTARES IN SIZE	AT LEAST 1 MILLION HECTARES IN SIZE
AFRICA	**727**	**149,541**	**4.9**	**301**	**91,639**	**426**	**57,902**	**26.1**	**4.7**
Algeria	19	11,919	5.0	12	11,801	7	118	10.5	10.5
Angola	5	2,641	2.1	1	790	4	1,851	60.0	0.0
Benin	2	778	6.9	2	778	0	0	100.0	0.0
Botswana	9	10,663	18.3	5	9,731	4	932	88.9	33.3
Burkina Faso	12	2,662	9.7	3	489	9	2,173	41.7	8.3
Burundi	3	89	3.2	0	0	3	89	0.0	0.0
Cameroon	14	2,050	4.3	7	1,032	7	1,019	57.1	0.0
Central African Republic	13	6,106	9.8	5	3,188	8	2,918	92.3	23.1
Chad	9	11,494	9.0	2	414	7	11,080	100.0	22.2
Congo	10	1,177	3.4	1	127	9	1,051	30.0	0.0
Egypt	12	793	0.8	4	99	8	695	8.3	0.0
Equatorial Guinea	0	0	0.0	0	0	0	0	0.0	0.0
Eritrea	0	0	0.0	0	0	0	0	0.0	0.0
Ethiopia	23	6,023	5.5	12	3,040	11	2,982	69.6	0.0
Gabon	6	1,045	3.9	1	15	5	1,030	33.3	0.0
Gambia	5	23	2.0	3	18	2	5	0.0	0.0
Ghana	9	1,104	4.6	7	1,097	2	7	33.3	0.0
Guinea	3	164	0.7	3	164	0	0	33.3	0.0
Guinea-Bissau	0	0	0.0	0	0	0	0	0.0	0.0
Ivory Coast	12	1,993	6.2	10	1,891	2	102	33.3	8.3
Kenya	36	3,504	6.0	32	3,451	4	52	19.4	2.8
Lesotho	1	7	0.2	0	0	1	7	0.0	0.0
Liberia	1	129	1.3	1	129	0	0	100.0	0.0
Libya	6	173	0.1	3	51	3	122	0.0	0.0
Madagascar	36	1,115	1.9	16	740	20	375	2.8	0.0
Malawi	9	1,059	8.9	5	696	4	362	33.3	0.0
Mali	11	4,012	3.2	1	350	10	3,662	54.5	18.2
Mauritania	4	1,746	1.7	3	1,496	1	250	75.0	25.0
Mauritius	1	4	1.8	0	0	1	4	0.0	0.0
Morocco	11	369	0.8	6	62	5	307	9.1	0.0
Mozambique	1	2	0.0	0	0	1	2	0.0	0.0
Namibia	12	10,218	12.4	6	9,000	6	1,218	50.0	25.0
Niger	5	8,416	6.6	1	220	4	8,196	60.0	20.0
Nigeria	19	2,971	3.2	6	2,226	13	745	47.4	0.0
Rwanda	2	327	12.4	2	327	0	0	50.0	0.0
Senegal	9	2,180	11.1	5	1,012	4	1,168	33.3	0.0
Sierra Leone	2	82	1.1	2	82	0	0	0.0	0.0
Somalia	1	180	0.3	0	0	1	180	100.0	0.0
South Africa	238	6,970	5.7	55	4,280	183	2,689	2.5	0.4
Sudan	16	9,383	3.7	9	8,514	7	869	43.8	25.0
Swaziland	3	40	2.3	0	0	3	40	0.0	0.0
Tanzania	31	13,936	14.7	12	4,100	19	9,836	61.3	9.7
Togo	11	647	11.4	3	357	8	290	27.3	0.0

(Continued on next page)

	ALL PROTECTED AREAS			TOTALLY PROTECTED AREAS		PARTIALLY PROTECTED AREAS		PERCENT OF PROTECTED AREAS	
	NUMBER	AREA (thousands of hectares)	% OF LAND AREA	NUMBER	AREA (thousands of hectares)	NUMBER	AREA (thousands of hectares)	AT LEAST 100,000 HECTARES IN SIZE	AT LEAST 1 MILLION HECTARES IN SIZE
Tunisia	6	44	0.3	6	44	0	0	0.0	0.0
Uganda	31	1,909	8.1	7	876	24	1,033	19.4	0.0
Zaire	8	9,917	4.2	8	9,917	0	0	100.0	50.0
Zambia	21	6,364	8.5	21	6,364	0	0	52.4	4.8
Zimbabwe	25	3,068	7.9	11	2,704	14	364	24.0	4.0
NORTH AND MIDDLE AMERICA	**2,549**	**230,199**	**10.2**	**1,297**	**112,100**	**1,252**	**118,098**	**10.7**	**1.8**
Belize	13	323	14.1	6	160	7	163	15.4	0.0
Canada	627	82,358	8.3	347	34,530	280	47,820	14.7	2.7
Costa Rica	28	648	12.7	16	503	12	144	3.6	0.0
Cuba	56	1,154	10.4	18	157	38	997	3.6	0.0
Dominican Republic	17	1,048	21.5	8	564	9	484	17.6	0.0
El Salvador	2	5	0.2	1	3	1	2	0.0	0.0
Guatemala	18	1,333	12.2	12	1,279	6	54	22.2	0.0
Haiti	3	10	0.3	2	8	1	2	0.0	0.0
Honduras	43	862	7.7	15	469	28	393	4.7	0.0
Jamaica	1	2	0.1	1	2	0	0	0.0	0.0
Mexico	68	9,854	5.0	41	1,925	27	7,929	16.2	2.9
Nicaragua	59	903	6.9	6	389	53	514	1.7	0.0
Panama	14	1,326	17.6	13	1,324	1	2	28.6	0.0
Trinidad and Tobago	5	16	3.0	1	2	4	14	0.0	0.0
United States	1,585	130,209	13.3	803	70,639	782	59,570	9.5	1.8
SOUTH AMERICA	**706**	**112,834**	**6.3**	**391**	**67,506**	**315**	**45,328**	**25.1**	**3.4**
Argentina	84	4,372	1.6	65	3,024	19	1,347	14.3	0.0
Bolivia	25	9,233	8.4	8	3,774	17	5,459	56.0	16.0
Brazil	272	32,189	3.8	149	20,423	123	11,766	23.5	1.8
Chile	66	13,725	18.1	32	8,375	34	5,350	25.8	7.6
Colombia	80	9,381	8.2	41	9,036	39	345	22.5	2.5
Ecuador	15	11,114	39.2	10	3,087	5	8,027	46.7	6.7
Guyana	1	59	0.3	1	59	0	0	0.0	0.0
Paraguay	20	1,495	3.7	13	1,368	7	127	10.0	0.0
Peru	22	4,176	3.2	15	4,044	7	133	27.3	9.1
Suriname	13	736	4.5	2	87	11	649	15.4	0.0
Uruguay	8	32	0.2	2	15	6	17	0.0	0.0
Venezuela	100	26,322	28.9	53	14,215	47	12,108	35.0	5.0
ASIA	**1,774**	**141,793**	**4.4**	**548**	**42,525**	**0**	**0**	**11.6**	**1.1**
Afghanistan	6	218	0.3	1	41	5	177	0.0	0.0
Armenia	4	214	7.2	4	214	0	0	25.0	0.0
Azerbaijan	12	191	2.2	12	191	0	0	0.0	0.0
Bangladesh	8	97	0.7	0	0	8	97	0.0	0.0
Bhutan	9	966	20.6	5	725	4	241	33.3	0.0
Cambodia	20	2,998	16.6	7	871	13	2,127	55.0	0.0
China	463	58,082	6.1	4	128	459	57,954	10.4	1.9
Georgia	15	187	2.7	15	187	0	0	0.0	0.0

(Continued on next page)

	ALL PROTECTED AREAS			TOTALLY PROTECTED AREAS		PARTIALLY PROTECTED AREAS		PERCENT OF PROTECTED AREAS	
	NUMBER	AREA (thousands of hectares)	% OF LAND AREA	NUMBER	AREA (thousands of hectares)	NUMBER	AREA (thousands of hectares)	AT LEAST 100,000 HECTARES IN SIZE	AT LEAST 1 MILLION HECTARES IN SIZE
India	339	14,337	4.4	66	3,874	273	10,463	6.2	0.0
Indonesia	168	18,564	9.7	97	14,397	71	4,167	18.5	3.0
Iran	67	8,299	5.0	26	2,986	41	5,314	26.9	1.5
Iraq	0	0	0.0	0	0	0	0	0	0
Israel	15	308	14.6	1	3	14	305	6.7	0.0
Japan	80	2,758	7.3	37	1,514	43	1,245	10.0	0.0
Jordan	10	290	3.3	1	1	9	289	10.0	0.0
Kazakhstan	20	988	0.4	9	892	11	96	10.0	0.0
Korea, North	2	58	0.5	1	44	1	14	0.0	0.0
Korea, South	27	693	7.0	5	19	22	674	3.7	0.0
Kuwait	2	27	1.5	1	2	1	25	0.0	0.0
Kyrgyzstan	5	284	1.4	5	284	0	0	20.0	0.0
Laos	0	0	0.0	0	0	0	0	0.0	0.0
Lebanon	1	4	0.3	1	4	0	0	0.0	0.0
Malaysia	51	1,484	4.5	41	903	10	581	9.8	0.0
Mongolia	15	6,168	3.9	14	5,618	1	550	13.3	6.7
Myanmar	2	173	0.3	1	161	1	13	50.0	0.0
Nepal	12	1,109	7.9	8	1,014	4	94	33.3	0.0
Oman	28	986	4.6	1	46	27	940	7.1	0.0
Pakistan	55	3,721	4.7	6	882	49	2,839	20.0	0.0
Philippines	27	606	2.0	15	267	12	339	3.7	0.0
Saudi Arabia	10	6,201	2.9	2	279	8	5,922	80.0	30.0
Singapore	1	3	4.5	0	0	1	3	0.0	0.0
Sri Lanka	56	796	12.1	25	468	31	328	0.0	0.0
Syria	0	0	0.0	0	0	0	0	0.0	0.0
Tajikistan	3	86	0.6	3	86	0	0	0.0	0.0
Thailand	111	7,020	13.7	74	4,336	37	2,684	18.9	0.0
Turkey	49	1,071	1.4	23	417	26	655	2.0	0.0
Turkmenistan	8	1,112	2.3	8	1,112	0	0	25.0	0.0
United Arab Emirates	0	0	0.0	0	0	0	0	0.0	0.0
Uzbekistan	10	244	0.5	10	244	0	0	0.0	0.0
Vietnam	52	1,334	4.0	9	202	43	1,131	1.9	0.0
Yemen	0	0	0.0	0	0	0	0	0	0
EUROPE	**2,923**	**223,905**	**8.9**	**565**	**157,432**	**2,358**	**66,473**	**7.2**	**0.5**
Albania	11	34	1.2	6	10	5	24	0.0	0.0
Austria	170	2,081	24.8	1	76	169	2,005	0.6	0.0
Belarus	11	265	1.3	2	144	9	120	0.0	0.0
Belgium	3	77	2.5	0	0	3	77	0.0	0.0
Bosnia-Herzegovina	5	25	0.5	1	17	4	8	0.0	0.0
Bulgaria	46	370	3.3	31	288	15	82	2.2	0.0
Croatia	30	392	6.9	11	74	19	318	3.3	0.0
Czech Republic	34	1,067	13.5	6	88	28	979	2.9	0.0
Denmark	114	99,618	44.9	10	98,278	104	1,340	3.5	1.8
Estonia	38	412	9.1	8	228	30	184	206	0.0
Finland	81	2,744	8.1	36	560	45	2,184	11.1	0.0
France	102	5,598	10.2	14	332	88	5,266	20.6	0.0
Germany	497	9,193	25.8	3	37	494	9,156	5.0	0.0

(Continued on next page)

	ALL PROTECTED AREAS			TOTALLY PROTECTED AREAS		PARTIALLY PROTECTED AREAS		PERCENT OF PROTECTED AREAS	
	NUMBER	AREA (thousands of hectares)	% OF LAND AREA	NUMBER	AREA (thousands of hectares)	NUMBER	AREA (thousands of hectares)	AT LEAST 100,000 HECTARES IN SIZE	AT LEAST 1 MILLION HECTARES IN SIZE
Greece	21	221	1.7	10	78	11	143	0.0	0.0
Hungary	53	574	6.2	5	159	48	415	0.0	0.0
Iceland	20	916	8.9	8	219	12	697	10.0	0.0
Ireland	11	47	0.7	5	37	6	10	0.0	0.0
Italy	171	2,275	7.5	12	473	159	1,801	1.2	0.0
Latvia	45	775	12.0	5	41	40	734	2.2	0.0
Lithuania	76	625	9.6	9	144	67	481	0.0	0.0
Macedonia	16	217	8.4	8	156	8	61	0.0	0.0
Moldova	3	12	0.4	3	12	0	0	0.0	0.0
Netherlands	85	429	1.5	38	292	47	137	1.2	0.0
Norway	113	5,536	17.1	75	5,054	38	482	9.7	0.9
Poland	111	3,069	9.8	16	155	95	2,914	4.5	0.0
Portugal	24	583	6.3	4	37	20	546	4.2	0.0
Romania	39	1,074	4.5	23	891	16	183	2.6	0.0
Russia	209	70,536	4.1	111	47,166	98	23,371	43.5	5.7
Slovakia	40	1,016	207	7	202	33	813	2.5	0.0
Slovenia	10	108	5.3	1	85	9	23	0.0	0.0
Spain	214	4,246	8.4	10	132	204	4,114	3.7	0.0
Sweden	197	2,982	6.6	52	1,444	145	1,538	3.0	0.0
Switzerland	109	731	17.7	1	17	108	714	0.0	0.0
Ukraine	19	485	0.8	15	312	4	173	0.0	0.0
United Kingdom	168	5,109	20.9	8	32	160	5,078	8.9	0.0
Yugoslavia	21	347	3.4	9	151	12	195	0.0	0.0
OCEANIA	**1,087**	**100,282**	**11.7**	**701**	**38,361**	**386**	**61,920**	**9.8**	**1.4**
Australia	889	94,077	12.2	568	32,459	321	61,618	10.3	1.6
Fiji	5	19	1.0	5	19	0	0	0.0	0.0
New Zealand	182	6,067	22.4	122	5,853	60	214	7.7	0.5
Papua New Guinea	5	82	0.2	3	7	2	75	0.0	0.0
Solomon Islands	0	0	0.0	0	0	0	0	0.0	0.0

Source: World Resources 1994–95.

Part VI

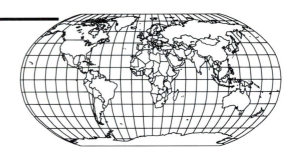

Human Impact on the Land

Map 43 The Risks of Desertification

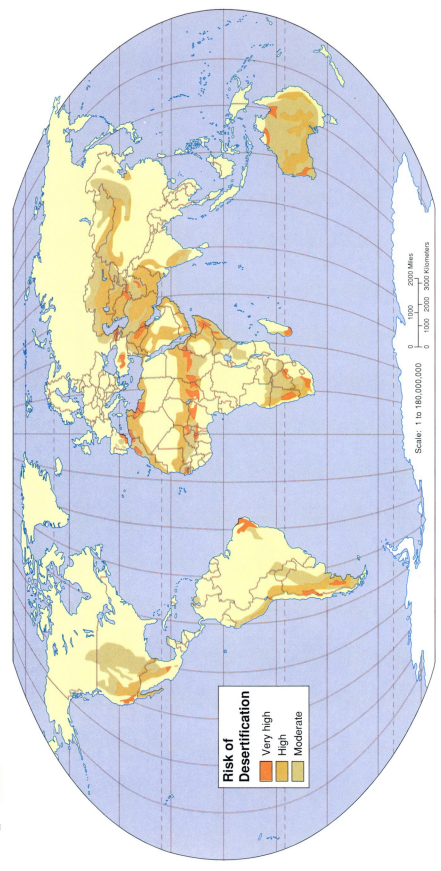

Risk of Desertification

- Very high
- High
- Moderate

Scale: 1 to 180,000,000

0   1000   2000 Miles

0   1000   2000   3000 Kilometers

The awkward-sounding term "desertification" refers to a reduction in the food-producing capacity of drylands through vegetation, soil, and water changes that culminate in either a drier climate or in soil and plant systems that are less efficient in their use of water. Most of the world's existing drylands—the shortgrass steppes, the tropical savannas, the bunchgrass regions of the desert fringe—are fairly intensively used for agriculture and are, therefore, subject to the kinds of pressures that culminate in desertification. Most desertification is a natural process that occurs near the margins of desert regions. It is caused by dehydration of the soil's surface layers during periods of drought and by high water loss through evaporation in an environment of high temperature and high winds. This natural process

is greatly enhanced by human agricultural activities that expose topsoil to wind and water erosion. Among the most important practices that cause desertification are (1) overgrazing of rangelands, resulting from too many livestock on too small an area of land; (2) improper management of soil and water resources in irrigation agriculture, leading to accelerated erosion and to salt buildup in the soil; (3) cultivation of marginal terrain with soils and slopes that are unsuitable for farming; (4) surface disturbances of vegetation (clearing of thorn scrub, mesquite, chaparral, and similar vegetation) without soil protection efforts being made or replanting being done; and (5) soil compaction by agricultural implements, domesticated livestock, and rain falling on an exposed surface.

Map 44 Global Soil Degradation

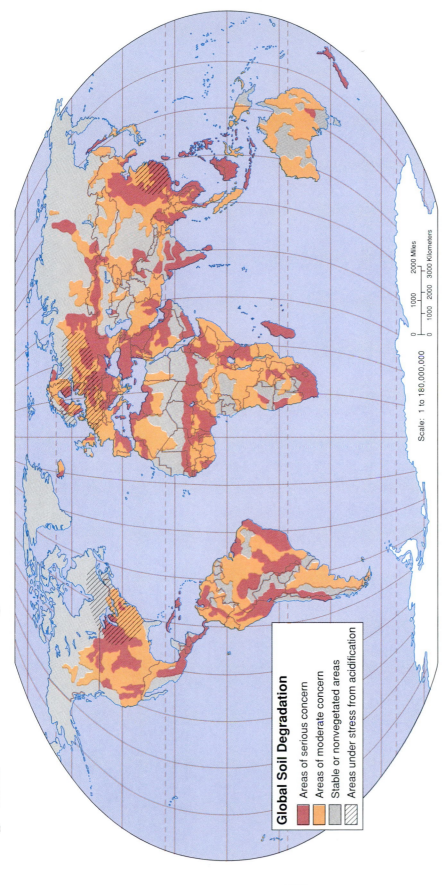

Global Soil Degradation

- Areas of serious concern
- Areas of moderate concern
- Stable or nonvegetated areas
- Areas under stress from acidification

Scale: 1 to 180,000,000

0   1000   2000 Miles
0   1000   2000   3000 Kilometers

Recent research has shown that more than 3 billion acres of the world's surface suffer from serious soil degradation, with more than 22 million acres so severely eroded or poisoned with chemicals that they can no longer support productive crop agriculture. Most of this soil damage has been caused by poor farming practices, overgrazing of domestic livestock, and deforestation. These activities strip away the protective cover of natural vegetation forests and grasslands, allowing wind and water erosion to remove the topsoil that contains necessary nutrients and soil microbes for plant growth. But millions of acres of topsoil have been degraded by chemicals as well. In some instances these chemicals are the result of overapplication of fertilizers, herbicides, pesticides, and other agricultural chemicals. In other instances, chemical deposition from industrial and urban wastes and from acid precipitation has poisoned millions of acres of soil. As the map shows, soil erosion and pollution are not problems just in developing countries with high population densities and increasing use of marginal lands. They also afflict the more highly developed regions of mechanized, industrial agriculture. While many methods for preventing or reducing soil degradation exist, they are seldom used because of ignorance, cost, or perceived economic inefficiency.

Map 45 Advancing Desert: Northern Africa, 1980–1994

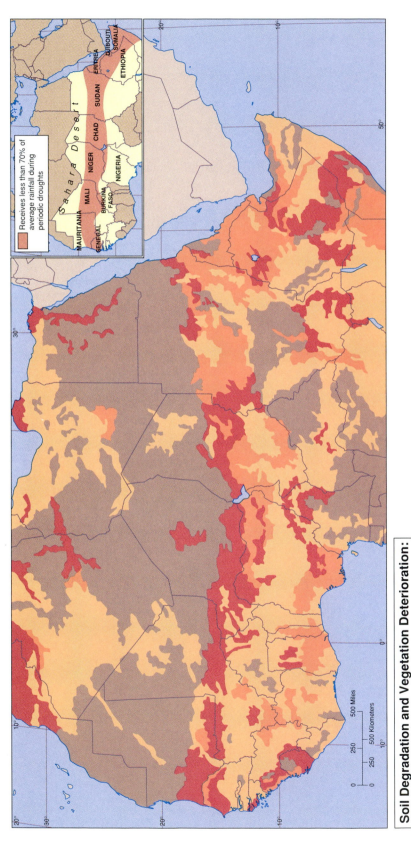

Receives less than 70% of average rainfall during periodic droughts

Soil Degradation and Vegetation Deterioration: The Sahel and Northern Africa

- No major soil/vegetation change
- Slow change
- Moderate change
- Rapid change

Desertification involves the conversion of useful dryland to virtual desert. Perhaps the best-documented process of desertification is that occurring in northern Africa, most particularly in the Sahel region, which lies between the Sahara and the humid forests of West and central Africa. Throughout northern Africa, the combination of a growing human and animal population with naturally occurring drought cycles has placed enormous pressures on the environment in the form of overgrazing and shortened vegetation rotation cycles. As a result, grasslands have diminished and thorn brush vege-

tation has increased. Thorn and scrub are much less capable of retaining water than grasslands, and more of the scanty precipitation runs off the scrub-covered surface than off grass cover. This precipitation carries off significant amounts of topsoil and leaves the area with less soil moisture to produce new vegetation or, through evapotranspiration, to serve as a source of atmospheric moisture from which new precipitation may result. The consequence is worsening drought and, eventually, the creation of desert from what was productive dryland.

Map 46 Environmental Degradation: Eastern Europe

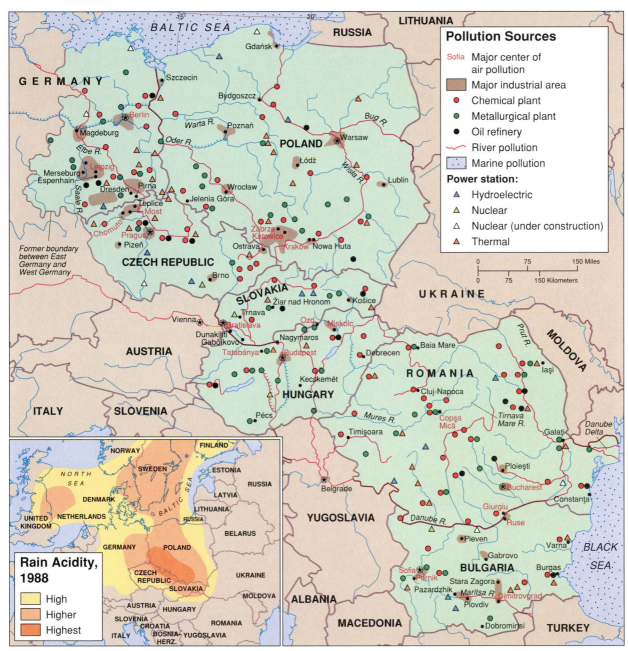

Pollution Sources

Sofia   Major center of air pollution

▨   Major industrial area

●   Chemical plant

●   Metallurgical plant

●   Oil refinery

◈   River pollution

▨   Marine pollution

Power station:

▲   Hydroelectric

▲   Nuclear

△   Nuclear (under construction)

▲   Thermal

Rain Acidity, 1988

☐ High

☐ Higher

☐ Highest

By almost any measure, eastern Europe is the world's most severely polluted region. The reasons for this include a Marxist economic approach that placed greater value on production than on environmental protection; the need to produce goods cheaply without the production costs associated with pollution controls; and environmental factors such as climate, terrain, hydrology, and available energy resources. Examples of environmental problems in this region are numerous. In the Czech Republic, half of the drinking water fails to meet the country's own environmental standards. In Bulgaria, industrial waste pollutes 70 percent of the country's farmland and 65 percent of its water supply. The capital of Romania has no effective sewage treatment plant, and most existing Romanian treatment facilities in the country do not work properly. To make matters worse, eastern Europe is the center of acid precipitation in the developed world, with winds bringing emissions from as far away as Great Britain and Norway. More than 1.5 million acres of woodland in the Czech Republic and Poland have been severely damaged by acid precipitation. The transition of eastern Europe to a market economy and ties with the European Union may allow this beleaguered region to begin cleaning up its environment. For many eastern Europeans whose health has already suffered, however, such a cleanup is too late.

Map 47 Soil Erosion: The United States Example

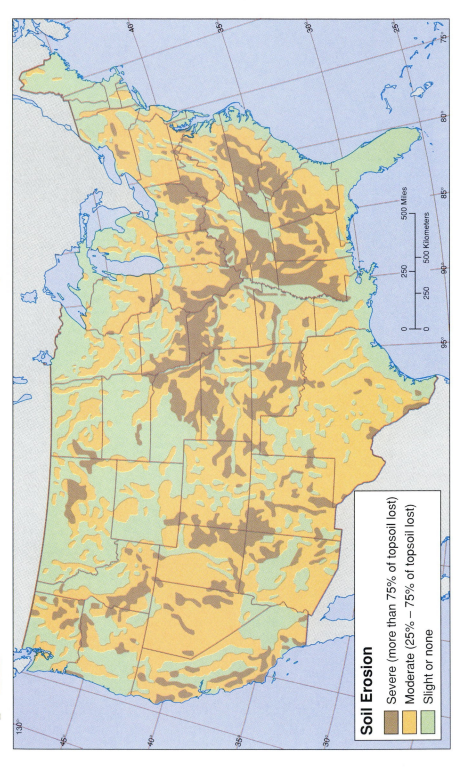

Soil Erosion

Severe (more than 75% of topsoil lost)

Moderate (25% – 75% of topsoil lost)

Slight or none

500 Miles

500 Kilometers

Erosion itself is a natural process, part of a larger cycle of geologic change. Human activities, however, can greatly accelerate the erosion process. Approximately one-third of the original topsoil on croplands in the United States has already been lost to water and wind erosion. And surveys show that the United States is still losing topsoil at a rate about seven times greater than the rate of natural soil formation. Enough topsoil erodes each day in the United States to fill railroad cars for a train 2,500 miles long, and two-thirds of this enormous quantity comes from less than 25 percent of the country's cropland. Just the nutrient loss alone from erosion involves a cost of about $18 billion per year. More costs are incurred (about $4

billion) when silt, plant nutrients, pesticides, and other agricultural chemicals are transported from farm fields into streams, ponds, lakes, and reservoirs. As grim as this picture appears, it could be worse (and, indeed, it is worse in countries like Russia and China, where little money for soil conservation exists). Of the world's major agricultural countries, only the United States is effectively reducing some of its soil losses. Yet effective soil conservation techniques are applied to only about half of all U.S. farmland and to less than half of the most erodible land. The map shows the tremendous cumulative erosion of lands in the agricultural heartland of the country, from which 75 percent of the topsoil has already been lost.

Map  48 Degrees of Human Disturbance

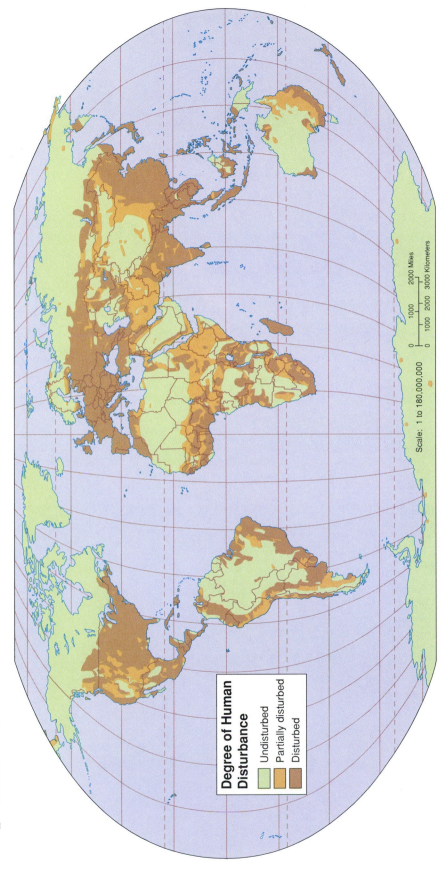

Degree of Human Disturbance

- Undisturbed
- Partially disturbed
- Disturbed

Scale: 1 to 180,000,000

| | 1000 | 2000 Miles |
| 0 | 1000 2000 | 3000 Kilometers |

All maps are generalizations, or models of the real world. This map is perhaps the culmination of all the generalizations made in the preceding pages of the atlas. The data on human disturbance have been gathered from a wide variety of sources, some of them conflicting and not all of them reliable. Nevertheless, at a global scale this map fairly depicts the state of the world in terms of the degree to which humans have modified its surface. The undisturbed areas, covered with natural vegetation, generally have population densities under 10 persons per square mile. These areas are, for the most part, in the most inhospitable parts of the world: too high, too dry, too cold for permanent human habitation in large numbers. The partially disturbed areas are normally agricultural areas, either subsistence (such as shifting cultivation) or extensive (such as livestock grazing).

They often contain areas of secondary vegetation, regrown after removal of original vegetation by humans. They are also often marked by a density of livestock in excess of carrying capacity, leading to overgrazing, which further alters the condition of the vegetation. The disturbed areas are those of permanent and intensive agriculture and urban settlement. The primary vegetation of these regions has been removed, with no evidence of regrowth or with current vegetation that is quite different from natural (potential) vegetation. Soils are in a state of depletion and degradation, and, in drier lands, desertification is a factor of human occupation. The disturbed areas match closely those areas of the world with the densest human populations.

Table N

Land Area and Use (in thousands of hectares)

	LAND AREA	POPULATION DENSITY (per 1,000 hectares)	DOMESTICATED LAND AS A % OF LAND AREA	CROPLAND 1991–1993	CROPLAND % CHANGE SINCE 1981–1983	PERMANENT PASTURE 1991–1993	PERMANENT PASTURE % CHANGE SINCE 1981–1983	FOREST AND WOODLAND 1991–1993	FOREST AND WOODLAND % CHANGE SINCE 1981–1983	OTHER LAND 1991–1993	OTHER LAND % CHANGE SINCE 1981–1983
AFRICA	**2,963,611**	**246**	**35**	**187,357**	**5.8**	**853,049**	**1.2**	**760,576**	**-3.1**	**1,162,630**	**-0.3**
Algeria	238,174	117	16	7,938	7.0	30,752	-3.2	3,969	-9.5	195,516	-0.5
Angola	124,670	89	26	3,483	2.5	29,000	0.0	51,917	-3.1	40,270	-3.9
Benin	11,062	489	21	1,877	4.1	442	0.0	3,407	-12.0	5,337	-7.3
Botswana	56,673	26	46	420	5.0	25,600	0.0	26,500	0.0	4,153	0.5
Burkina Faso	27,360	377	35	3,564	23.5	6,000	0.0	13,800	0.0	3,996	0.5
Burundi	2,568	2,489	89	1,357	3.9	915	0.5	85	0.0	211	17.0
Cameroon	46,540	284	19	7,033	1.2	2,000	0.0	35,900	0.0	1,607	26.3
Central African Republic	62,298	53	8	2,015	2.9	3,000	0.0	46,700	0.0	10,583	5.2
Chad	125,920	51	38	3,239	2.8	45,000	0.0	32,400	0.0	45,281	0.5
Congo	34,150	76	30	170	11.6	10,000	0.0	21,120	-0.9	2,860	0.2
Egypt	99,545	632	8	2,760	11.7	4,934	12.7	31	0.0	91,820	-6.4
Equatorial Guinea	2,805	143	12	230	0.0	104	0.0	1,297	0.1	1,174	0.9
Eritrea	10,100	350	60	1,280	X	1,600	X	2,000	X	5,220	0.1
Ethiopia	110,100	500	53	13,930	0.0	44,825	-1.0	26,950	-3.4	24,395	93.5
Gabon	25,767	51	20	459	1.5	4,700	0.0	19,900	-0.5	708	-5.8
Gambia	1,000	1,118	27	180	13.0	90	0.0	280	0.0	450	-12.7
Ghana	22,754	767	41	4,320	23.4	5,000	0.0	7,943	-8.0	5,491	4.6
Guinea	24,572	273	25	730	2.4	5,500	0.0	14,480	-3.9	3,862	2.4
Guinea-Bissau	2,812	382	50	340	13.7	1,080	0.0	1,070	0.0	322	-14.6
Ivory Coast	31,800	448	53	3,703	16.7	13,000	0.0	7,080	-24.4	8,017	12.7
Kenya	56,914	497	45	4,517	5.5	21,300	0.0	16,800	0.0	14,297	-21.9
Lesotho	3,035	675	76	320	10.4	2,000	0.0	80	0.0	635	1.7
Liberia	9,675	314	63	374	0.9	5,700	0.0	1,707	-4.8	1,894	4.1
Libya	175,954	31	9	2,167	3.5	13,300	0.8	840	-13.2	159,647	-13.6
Madagascar	58,154	254	47	3,104	3.3	24,000	0.0	23,200	35.1	7,850	0.2
Malawi	9,408	1,183	38	1,697	22.1	1,840	0.0	3,700	0.0	2,171	1.3
Mali	122,019	88	27	2,270	10.6	30,000	0.0	6,907	-1.1	82,843	12.3
Mauritania	102,522	22	38	207	6.2	39,250	0.0	4,413	-3.9	58,652	-0.1
Mauritius	203	5,502	56	106	-0.9	7	0.0	44	-2.1	46	-0.1
Morocco	44,630	606	69	9,781	15.9	20,900	0.0	8,290	-24.1	5,659	-32.6
Mozambique	78,409	204	60	3,163	2.7	44,000	0.0	14,053	6.0	17,192	32.0
Namibia	82,329	19	47	662	0.5	38,000	0.0	18,030	-7.7	25,637	-6.3
Niger	126,670	72	10	3,605	1.5	8,913	-3.4	2,500	-1.8	111,652	-1.3
Nigeria	91,077	1,227	79	32,368	6.1	40,000	0.0	11,400	0.8	7,309	-0.2
Rwanda	2,467	3,223	66	1,167	8.5	453	-15.2	550	-20.3	297	-14.1
Senegal	19,253	432	28	2,350	0.0	3,100	0.0	10,467	-4.8	3,336	-5.8
Sierra Leone	7,162	630	38	540	5.6	2,203	-0.1	2,043	-4.4	2,376	-14.5
Somalia	62,734	147	70	1,032	2.2	43,000	0.0	16,000	-2.9	2,702	-1.4
South Africa	122,104	340	77	13,177	-0.1	81,378	-0.0	8,200	6.7	19,349	37.8
Sudan	237,600	118	52	12,950	3.6	110,000	12.2	44,340	-6.1	70,310	13.6
Swaziland	1,720	497	73	191	35.4	1,070	-6.5	118	15.7	341	-2.5

(Continued on next page)

	LAND AREA	POPULATION DENSITY (per 1,000 hectares)	DOMESTICATED LAND AS A % OF LAND AREA	CROPLAND		PERMANENT PASTURE		FOREST AND WOODLAND		OTHER LAND	
				1991–1993	% CHANGE SINCE 1981–1983	1991–1993	% CHANGE SINCE 1981–1983	1991–1993	% CHANGE SINCE 1981–1983	1991–1993	% CHANGE SINCE 1981–1983
Tanzania	88,359	336	44	3,500	19.2	35,000	0.0	33,500	–14.4	16,359	–31.0
Togo	5,439	761	48	2,430	3.0	200	0.0	933	–8.5	1,876	–0.9
Tunisia	15,536	573	52	4,897	2.2	3,525	5.1	664	18.8	6,451	5.9
Uganda	19,965	1,067	43	6,763	13.4	1,800	0.0	5,503	–7.7	5,898	5.8
Zaire	226,705	194	10	7,893	2.9	15,000	0.0	173,860	–1.7	29,952	–9.6
Zambia	74,339	127	47	5,271	2.2	30,000	0.0	28,727	–2.3	10,341	–5.5
Zimbabwe	38,685	291	20	2,864	3.4	4,856	0.0	8,800	–7.4	22,165	–2.7
NORTH AND MIDDLE AMERICA	**2,178,176**	**209**	**29**	**271,300**	**–0.9**	**362,033**	**0.1**	**854,897**	**5.7**	**689,945**	**6.3**
Belize	2,280	94	5	57	7.5	48	9.1	2,100	0.0	75	10.7
Canada	922,097	32	8	45,523	–1.3	27,933	–3.3	494,000	11.6	354,640	14.0
Costa Rica	5,106	671	56	530	3.5	2,337	9.1	1,570	–5.2	670	19.1
Cuba	10,982	1,005	57	3,337	3.8	2,970	11.4	2,403	–9.2	2,273	8.0
Dominican Republic	4,838	1,617	30	1,449	1.4	2	0.0	608	–3.7	2,780	–0.1
El Salvador	2,072	2,784	65	731	0.8	610	0.0	104	–18.8	627	–2.9
Guatemala	10,843	980	41	1,817	2.3	2,534	91.8	5,271	17.4	1,221	166.6
Haiti	2,756	2,605	51	908	1.3	495	–2.3	140	0.0	1,212	–0.0
Honduras	11,189	505	32	1,904	7.7	1,511	0.7	6,000	0.0	1,774	8.3
Jamaica	1,083	2,259	44	219	–4.8	257	0.0	184	–4.5	423	–4.7
Mexico	190,869	491	52	24,727	0.2	74,499	0.0	48,700	4.1	42,943	4.6
Nicaragua	11,875	373	57	1,272	1.0	5,483	9.7	3,223	–24.3	1,896	–28.4
Panama	7,443	353	29	658	15.0	1,487	10.9	3,260	–18.0	2,038	–23.8
Trinidad and Tobago	513	2,546	26	121	3.4	11	0.0	235	3.1	146	7.6
United States	957,311	275	45	187,776	–1.2	239,172	–0.4	286,400	–2.0	243,963	–3.8
SOUTH AMERICA	**1,752,925**	**182**	**34**	**104,567**	**1.6**	**495,884**	**3.9**	**846,721**	**–4.1**	**305,753**	**–5.2**
Argentina	273,669	126	62	27,200	0.0	142,033	–0.7	50,900	0.0	53,536	–1.8
Bolivia	108,438	68	27	2,373	11.1	26,517	–1.7	58,000	0.0	21,549	–1.1
Brazil	845,651	191	28	50,560	0.3	185,767	6.8	488,833	–4.8	120,491	–10.3
Chile	74,880	190	24	4,293	0.3	13,583	3.7	16,500	0.0	40,504	1.2
Colombia	103,870	338	44	5,450	4.2	40,567	4.8	49,633	–5.8	8,220	–12.1
Ecuador	27,684	414	18	3,012	20.2	2,091	–0.0	15,600	0.6	6,981	8.7
Guyana	19,685	42	9	496	0.2	1,230	0.5	16,413	0.3	1,546	3.3
Paraguay	39,730	125	60	2,258	19.3	21,600	30.6	12,983	–32.5	2,888	–28.6
Peru	128,000	186	24	3,630	0.7	27,120	0.0	84,800	–0.1	12,450	–0.2
Suriname	15,600	27	1	68	24.4	21	6.8	15,000	0.8	511	26.4
Uruguay	17,481	182	85	1,304	–6.8	13,520	–0.6	930	0.0	1,727	–10.5
Venezuela	88,205	248	25	3,912	4.1	17,783	2.8	29,828	–8.2	36,682	–5.5
ASIA	**3,089,163**	**1,119**	**X**	**470,322**	**2.9**	**799,881**	**12.9**	**533,087**	**–1.9**	**958,376**	**9.1**
Afghanistan	65,209	309	58	8,054	0.0	30,000	0.0	1,900	0.0	25,255	0.0
Armenia	2,840	1,267	X	X	X	X	X	X	X	2,840	0.0
Azerbaijan	8,610	878	49	2,000	2.0	2,233	–4.1	960	–12.5	3,417	–5.7
Bangladesh	13,017	9,252	79	9,703	6.1	600	0.0	1,896	–12.2	818	36.3

(Continued on next page)

	LAND AREA	POPULATION DENSITY (per 1,000 hectares)	DOMESTICATED LAND AS A % OF LAND AREA	CROPLAND 1991–1993	CROPLAND % CHANGE SINCE 1981–1983	PERMANENT PASTURE 1991–1993	PERMANENT PASTURE % CHANGE SINCE 1981–1983	FOREST AND WOODLAND 1991–1993	FOREST AND WOODLAND % CHANGE SINCE 1981–1983	OTHER LAND 1991–1993	OTHER LAND % CHANGE SINCE 1981–1983
Bhutan	4,700	349	9	133	6.7	272	2.5	3,100	20.4	1,194	45.3
Cambodia	17,652	581	25	2,367	47.9	1,967	239.1	11,667	-11.3	1,652	40.0
China	932,641	1,310	X	95,975	-3.3	X	X	X	X	836,666	-0.4
Georgia	6,970	783	43	993	0.3	2,033	-9.8	2,717	-6.1	1,227	-32.2
India	297,319	3,147	61	169,547	0.6	11,533	-4.1	68,330	1.4	47,909	3.0
Indonesia	181,157	1,091	24	30,993	19.2	11,776	-1.3	111,258	-3.7	27,130	2.1
Iran	163,600	411	38	18,057	22.1	44,000	0.0	11,400	0.0	90,143	3.6
Iraq	43,737	468	22	5,450	0.1	4,000	0.0	192	0.0	34,095	0.0
Israel	2,062	2,730	28	435	4.8	145	12.7	124	12.7	1,358	3.7
Japan	37,652	3,322	14	4,511	-6.6	656	9.8	25,187	-0.0	7,299	-3.7
Jordan	8,893	612	13	404	18.4	791	0.1	70	6.6	7,628	0.9
Kazakhstan	266,980	64	83	35,328	-1.6	186,452	-1.0	9,600	-6.5	35,600	-8.6
Korea, North	12,041	1,986	17	2,010	4.5	50	0.0	2	0.0	9,979	0.9
Korea, South	9,873	4,557	22	2,072	-4.9	91	43.9	7,370	0.0	340	-23.1
Kuwait	1,782	868	8	5	150.0	137	2.0	6,464	-1.4	-4,824	1.8
Kyrgyzstan	19,130	248	54	1,387	(-3.9)	8,943	-1.7	2	0.0	8,798	-2.4
Laos	23,080	212	7	807	11.6	800	0.0	703	-11.3	20,770	-0.0
Lebanon	1,023	2,941	31	306	2.7	10	0.0	2,819	2.3	-2,112	-3.4
Malaysia	32,855	613	15	4,880	0.4	27	0.0	20,347	-1.8	7,601	-4.6
Mongolia	156,650	15	81	1,399	11.1	124,800	1.1	13,750	-9.4	16,701	0.6
Myanmar	65,755	708	16	10,061	-0.1	359	-0.7	32,397	0.9	22,938	1.2
Nepal	13,680	1,602	32	2,354	1.5	2,000	4.2	5,750	4.4	3,576	9.9
Oman	21,246	102	5	62	49.6	1,000	0.0	X	X	20,184	0.1
Pakistan	77,088	1,823	34	22,890	12.4	5,000	0.0	3,470	14.5	45,728	6.5
Philippines	29,817	2,267	35	9,177	3.8	1,277	17.1	13,600	13.4	5,764	37.0
Saudi Arabia	214,969	83	58	3,719	75.7	120,000	41.2	1,800	36.7	89,450	41.5
Singapore	61	46,689	X	1	-84.2	X	X	3	0.0	57	-9.4
Sri Lanka	6,463	2,840	36	1,903	2.1	439	0.1	2,126	21.4	1,995	20.7
Syria	18,378	798	75	5,770	0.8	8,018	-4.0	679	37.9	3,911	-2.5
Tajikistan	14,270	428	31	836	-7.2	3,507	1.1	535	7.8	9,391	0.1
Thailand	51,089	1,151	42	20,775	9.4	797	17.2	13,557	-13.8	15,960	-1.7
Turkey	76,963	805	52	27,583	0.6	12,378	22.6	20,199	0.0	16,803	14.6
Turkmenistan	48,810	84	77	1,462	-39.1	36,274	0.8	4,000	-10.4	7,074	-16.0
United Arab Emirates	8,360	228	3	39	36.0	200	0.0	3	0.0	8,118	0.1
Uzbekistan	42,540	537	63	4,728	4.6	22,183	-5.9	1,323	-42.5	14,306	-15.1
Vietnam	32,549	2,290	22	6,607	0.4	328	7.9	9,639	-8.8	15,975	-5.5
Yemen	52,797	275	33	1,481	1.1	16,065	0.0	2,000	-33.3	33,251	-3.0
EUROPE	**2,269,180**	**320**	**X**	**136,757**	**-2.7**	**80,794**	**-5.6**	**158,219**	**1.4**	**808,204**	**0.2**
Albania	2,740	1,256	41	702	-0.8	424	4.7	1,050	2.3	564	6.6
Austria	8,273	963	42	1,509	-5.6	1,978	-2.2	3,228	-1.0	1,557	-10.8
Belarus	20,760	488	45	6,255	-1.7	3,128	-6.2	7,009	-5.2	4,369	-16.0
Belgium	3,282	3,205	52	1,007	30.6	691	-9.4	700	0.4	885	18.9
Bosnia-Herzegovina	5,100	678	38	1,007	X	1,067	X	2,100	X	927	450.4
Bulgaria	11,055	793	55	4,267	2.8	1,878	-7.4	3,875	0.5	1,035	-1.5

(Continued on next page)

	LAND AREA	POPULATION DENSITY (per 1,000 hectares)	DOMESTICATED LAND AS A % OF LAND AREA	CROPLAND 1991–1993	CROPLAND % CHANGE SINCE 1981–1983	PERMANENT PASTURE 1991–1993	PERMANENT PASTURE % CHANGE SINCE 1981–1983	FOREST AND WOODLAND 1991–1993	FOREST AND WOODLAND % CHANGE SINCE 1981–1983	OTHER LAND 1991–1993	OTHER LAND % CHANGE SINCE 1981–1983
Croatia	5,592	804	43	1,415	–13.2	1,247	–21.2	2,081	2.3	849	–59.5
Czech Republic	7,728	1,332	54	3,293	X	873	X	2,629	X	933	728.3
Denmark	4,243	1,221	65	2,549	–3.3	206	–14.9	445	–9.7	1,043	–16.3
Estonia	4,227	362	34	1,146	15.2	312	–6.8	1,986	19.2	783	57.2
Finland	30,461	168	9	2,560	2.1	116	–23.6	23,198	–0.5	4,587	–2.3
France	55,010	1,054	55	19,297	1.6	11,023	–12.8	14,884	2.0	9,806	–10.4
Germany	34,927	2,336	50	12,009	–3.7	5,274	–11.0	10,700	4.0	6,943	–10.1
Greece	12,890	811	68	3,506	–11.2	5,253	–0.0	2,620	0.0	1,510	–29.4
Hungary	9,234	1,095	66	5,077	–4.2	1,165	–9.2	1,726	6.1	1,266	–19.2
Iceland	10,025	27	23	6	–20.8	2,274	0.0	120	0.0	7,625	–0.0
Ireland	6,889	516	81	926	–12.2	4,691	0.5	320	–0.8	951	–11.1
Italy	29,406	1,945	55	11,915	–3.7	4,284	–15.7	6,769	6.2	6,438	–13.2
Latvia	6,205	412	41	1,715	–1.3	822	9.0	12,507	–5.6	–8,839	7.8
Lithuania	4,551	813	77	3,043	–3.6	459	–15.1	1,980	1.0	–930	18.9
Macedonia	2,543	851	51	663	X	636	X	1,002	X	242	952.3
Moldova	3,297	1,344	79	2,202	–0.4	378	8.0	421	74.3	296	67.5
Netherlands	3,392	4,570	59	922	12.0	1,065	–10.1	343	15.9	1,061	2.4
Norway	30,683	141	3	888	5.8	120	18.4	8,330	0.0	21,344	0.3
Poland	30,442	1,261	61	14,694	–0.9	4,043	–0.6	8,779	0.9	2,926	–2.7
Portugal	9,195	1,068	44	3,169	0.7	839	0.1	3,293	12.3	1,894	20.2
Romania	23,034	991	64	9,974	–5.3	4,820	8.7	6,681	1.8	1,559	–3.5
Russia	1,699,580	86	12	133,141	–1.8	77,985	–6.8	778,512	1.9	709,941	0.9
Slovakia	4,808	1,113	51	1,623	X	825	X	1,990	X	370	1,198.3
Slovenia	2,012	967	43	302	X	559	X	1,018	X	133	1,412.8
Spain	49,944	793	60	19,897	–2.9	10,281	–2.7	15,970	2.4	3,796	–13.6
Sweden	41,162	213	8	2,779	–5.8	573	–18.9	28,000	0.1	9,809	–2.9
Switzerland	3,955	1,821	40	449	9.1	1,279	–20.5	1,185	12.7	1,042	–15.3
Ukraine	57,935	887	72	34,458	–2.9	7,471	6.6	10,278	36.4	5,728	37.9
United Kingdom	24,160	2,411	71	6,442	–7.7	11,112	–1.6	2,424	12.0	4,182	–11.0
Yugoslavia	X	X	X	7,730	–1.4	6,352	–0.6	9,120	–1.4	–23,202	1.2
OCEANIA	**845,349**	**34**	**57**	**51,619**	**4.3**	**430,738**	**–3.7**	**199,962**	**24.4**	**163,030**	**15.2**
Australia	764,444	24	60	46,579	3.8	416,567	–3.7	145,000	36.8	156,298	15.9
Fiji	1,827	429	24	257	42.8	174	34.1	1,185	0.0	211	57.6
New Zealand	26,799	133	65	3,831	9.4	13,577	–5.3	7,377	3.6	2,014	–8.3
Papua New Guinea	45,286	95	1	412	11.1	81	–17.3	42,000	0.0	2,793	0.9
Solomon Islands	2,799	135	3	57	8.2	39	0.0	2,450	–3.5	253	–33.9

Source: World Resources 1996–97.

Acknowledgments

The quality of this book was improved by the insights provided by the following reviewers:

Dianne Draper, University of Calgary
Gian Gupta, University of Maryland–Eastern Shore
Vishnu R. Khade, Eastern Connecticut State University

Sources

After the storm. (1991, August). *National Geographic, 180.*

Alaska's big spill. (1990, January). *National Geographic, 177.*

Amazonia [map]. (1994). *National Geographic, 186.*

An atmosphere of uncertainty. (1987, April). *National Geographic, 171.*

Crabb, C. (1993, January). Soiling the planet. *Discover, 14*(1), 74–75. [For information regarding Map 44 on page 109 of this book]

DeBlij, H. J., & Muller, P. (1994). *Geography: Realms, regions and concepts* (7th ed., revised). New York: John Wiley & Sons.

Department of Geography, Pennsylvania State University. (1996). Unpublished computer model output. State College, PA: Pennsylvania State University.

Domke, K. (1988). *War and the changing global system.* New Haven, CT: Yale University Press.

Eastern Europe's dark dawn. (1991, June). *National Geographic, 179.*

Economic consequences of the accident at Chernobyl nuclear plant. (1987). *PlanEcon Reports, 3.*

Environmental Protection Agency. (1996). Unpublished data [Online]. Available: http://www.epa.gov.

Fellman, J., Getis, A., & Getis, J. (1995). *Human geography: Landscapes of human activities* (4th ed.). Dubuque, IA: Wm. C. Brown Publishers.

Information please almanac 1996. (1995). Boston & New York: Houghton Mifflin.

Johnson, D. (1977). *Population, society, and desertification.* New York: United Nations Conference on Desertification, United Nations Environment Programme.

Kuchler, A. W. (1949). Natural vegetation. *Annals of the Association of American Geographers, 39.*

Küppen, W., & Geiger, R. (1954). *Klima der erde* [Climate of the earth]. Darmstadt, Germany: Justus Perthes.

Lindeman, M. (1990). *The United States and the Soviet Union: Choices for the 21st century.* Guilford, CT: Dushkin Publishing Group.

Mather, J. R. (1974). *Climatology: Fundamentals and applications.* New York: McGraw-Hill.

Miller, G. T. (1992). *Living in the environment* (7th ed.). Belmont, CA: Wadsworth. [For Map 32 on page 66 of this book. © 1992 by Wadsworth, Inc. Used by permission of the publisher.]

Murphy, R. E. (1968). Landforms of the world [Map supplement No. 9]. *Annals of the Association of American Geographers, 58*(1), 198–200.

National Aeronautics and Space Administration. (1994–1996). Unpublished data and images [Online]. Available: http://www.nasa.gov.

National Oceanic and Atmospheric Administration. (1996). Unpublished data [Online]. Available: http://www.noaa.gov.

The Oglalla Aquifer. (1993, March). *National Geographic, 183.*

Population Reference Bureau. (1994). *1994 world population data sheet.* New York: Population Reference Bureau.

A reconnaissance-level inventory of the amount of wilderness remaining in the world. (1989). *Ambio, 14.*

Rondônia: Brazil's imperiled rainforest. (1988, December). *National Geographic, 174.*

Rourke, J. T. (1995). *International politics on the world stage* (5th ed.). Guilford, CT: Dushkin Publishing Group/Brown & Benchmark Publishers.

Shelley, F., & Clarke, A. (1994). *Human and cultural geography: A global perspective.* Dubuque, IA: Wm. C. Brown Publishers.

Soiling the planet. (1993, January). *Discover, 14.*

Spector, L. S., & Smith, J. R. (1990). *Nuclear ambitions: The spread of nuclear weapons.* Boulder, CO: Westview Press.

This fragile earth [map]. (1988, December). *National Geographic, 174.*

Thornthwaite, C. W., & Mather, J. R. (1955). *The water balance* [Publications in Climatology No. 8]. Centerton, NJ: Drexel Institute of Technology, Laboratory of Climatology.

Times atlas of world history. (1978). Maplewood, NJ: Hammond.

United Nations Food and Agriculture Organization (FAO). (1995). *Forest resources assessment 1990: Global synthesis* [FAO Forestry Paper No. 124]. Rome: FAO.

United Nations Population Fund. (1992). *The state of the world's population.* New York: United Nations Population Fund.

United Nations Population Reference Bureau. (1990). *World development report.* New York: Oxford University Press.

U.S. Arms Control and Disarmament Agency. (1993). *World military expenditures and arms transfers.* Washington, DC: U.S. Government Printing Office.

U.S. Census Bureau. (1994). *World population profile.* Washington, DC: U.S. Government Printing Office.

U.S. Central Intelligence Agency. (1996). Unpublished data [Online]. Available: http://www.odci.gov/cia/publications.

U.S. Central Intelligence Agency. (1995). *World factbook 1995.* Available: http://www.odci.gov/cia/publications/95fact/index.html.

U.S. Central Intelligence Agency. (1996). *World factbook 1996–97.* Washington, DC: Brasseys.

U.S. Department of Energy. (1996). *U.S.–Canada memorandum of intent on transboundary air pollution.* Washington, DC: U.S. Government Printing Office.

U.S. Forest Service. (1989). *Ecoregions of the continents.* Washington, DC: U.S. Government Printing Office.

U.S. Soil Conservation Service [now the U.S. Natural Resources Conservation Service]. (1996). Unpublished data. Washington, DC: U.S. Soil Conversation Service.

The world almanac and book of facts 1996. (1995). Mahwah, NJ: World Almanac Books.

World Bank. (1995). *World development report 1995.* Geneva: World Bank.

World Conservation Monitoring Centre. (1996). Unpublished data. Cambridge, England: World Conservation Monitoring Centre.

World Health Organization. (1994). *World health statistics annual.* Geneva: World Health Organization.

World Resources Institute. (1996). *World resources 1996–97.* New York: Oxford University Press.

WorldWatch Institute. (1987). *Reassessing nuclear power: The fallout from Chernobyl* [WorldWatch paper no. 75]. New York: WorldWatch Institute. [For Map 23 on page 53 of this book. © 1987 by WorldWatch Institute, Washington, DC. Used by permission.]